Integrating Landscape Approaches and Multi-Resource Analysis into Natural Resource Management

SUMMARY OF A WORKSHOP

Dominic A. Brose, *Rapporteur*

Committee on the Practice of Sustainability Science

Science and Technology for Sustainability Program

Policy and Global Affairs

The National Academies of
SCIENCES · ENGINEERING · MEDICINE

THE NATIONAL ACADEMIES PRESS
Washington, DC
www.nap.edu

THE NATIONAL ACADEMIES PRESS 500 Fifth Street, NW Washington, DC 20001

This summary report and the workshop on which it was based were supported by the U.S. Geological Survey under Grant/Cooperative Agreement Number G11AC20535 and the National Oceanic and Atmospheric Administration under contract number WC133R-11-CQ0048, TO #5. Any opinions, findings, conclusions, or recommendations expressed in this publication do not necessarily reflect the views of any organization or agency that provided support for the project.

International Standard Book Number 13: 978-0-309-39215-0
International Standard Book Number 10: 0-309-39215-2
Digital Object Identifier: 10.17226/21917

Additional copies of this report are available from the National Academies Press, 500 Fifth Street, NW, Keck 360, Washington, DC 20001; (800) 624-6242 or (202) 334-3313; http://www.nap.edu.

Printed in the United States of America

Suggested citation: National Academies of Sciences, Engineering, and Medicine. 2016. *Integrating Landscape Approaches and Multi-Resource Analysis into Natural Resource Management*. Washington, DC: The National Academies Press. doi: 10.17226/21917.

The National Academies of
SCIENCES · ENGINEERING · MEDICINE

The **National Academy of Sciences** was established in 1863 by an Act of Congress, signed by President Lincoln, as a private, nongovernmental institution to advise the nation on issues related to science and technology. Members are elected by their peers for outstanding contributions to research. Dr. Ralph J. Cicerone is president.

The **National Academy of Engineering** was established in 1964 under the charter of the National Academy of Sciences to bring the practices of engineering to advising the nation. Members are elected by their peers for extraordinary contributions to engineering. Dr. C. D. Mote, Jr., is president.

The **National Academy of Medicine** (formerly the Institute of Medicine) was established in 1970 under the charter of the National Academy of Sciences to advise the nation on medical and health issues. Members are elected by their peers for distinguished contributions to medicine and health. Dr. Victor J. Dzau is president.

The three Academies work together as the **National Academies of Sciences, Engineering, and Medicine** to provide independent, objective analysis and advice to the nation and conduct other activities to solve complex problems and inform public policy decisions. The Academies also encourage education and research, recognize outstanding contributions to knowledge, and increase public understanding in matters of science, engineering, and medicine.

Learn more about the National Academies of Sciences, Engineering, and Medicine at **www.national-academies.org**.

COMMITTEE ON THE PRACTICE OF SUSTAINABILITY SCIENCE

Ingrid C. Burke (Chair), Director, Haub School of Environment and Natural Resources, University of Wyoming and Ruckelshaus Institute
Steve Bergman, Principal Regional Geologist, Global Geology Research Team, Shell International Exploration & Production Co.
Robert J. Johnston, Director, George Perkins Marsh Institute and Professor of Economics, Clark University
Joseph Kiesecker, Lead Scientist, The Nature Conservancy's Conservation Lands Team
Elizabeth Murray, Research Biologist, Wetlands and Coastal Ecology, U.S. Army Engineer Research and Development Center

Staff

Jerry Miller, Director, Science and Technology for Sustainability Program
Dominic Brose, Program Officer, Science and Technology for Sustainability Program
Jennifer Saunders, Senior Program Officer, Science and Technology for Sustainability Program
Emi Kameyama, Program Associate, Science and Technology for Sustainability Program
Yasmin Romitti, Research Assistant, Science and Technology for Sustainability Program

Preface and Acknowledgments

The responsible management of natural resources for present-day needs and future generations requires integrated approaches that are place-based, embrace systems thinking, and incorporate the social, economic, and environmental considerations of sustainability. Landscape-scale analysis takes a holistic view by focusing on the spatial scales most appropriate for the resource values being managed. Multi-resource analysis is an approach to landscape-scale analysis that integrates information among multiple natural resources, including ecosystem services, and is designed to evaluate impacts and tradeoffs between development and conservation at landscape scales. This approach implicitly addresses the social, economic, and ecological functional relationships.

The Science and Technology for Sustainability Program of the National Academies of Sciences, Engineering, and Medicine convened a committee to plan and hold a workshop on using landscape and multi-resource analyses to better inform federal decision making for the sustainable management of natural resources. The U.S. Geological Survey (USGS) and National Oceanic and Atmospheric Administration (NOAA) sponsored the workshop and are just two of several federal agencies embracing these approaches to encourage broader interdisciplinary thinking across and within their mission areas.

The goal of the workshop was to identify ways to better integrate landscape and multi-resource analyses across several focus areas, including adaptive management, ecosystem services, and resilience, which together form a scientific foundation for making sustainable natural resource management decisions. The workshop used case studies of pragmatic approaches that aim to integrate landscape and multi-resource analyses into practice on issues related to sustainable natural resource management. The workshop was organized around discussions of knowledge gaps and priority areas for research. The workshop was held on June 2, 2015, in Washington, D.C. A participatory approach was used that allowed for open discussion, and included participants from federal agencies, policy makers, and the broader scientific community.

This workshop summary was prepared by the workshop rapporteur as a factual summary of what was presented and discussed at the workshop. The planning committee's role was limited to planning and convening the workshop. The statements made are those of the rapporteur and do not necessarily represent positions of the workshop participants as a whole, the planning committee, or the National Academies of Sciences, Engineering, and Medicine. I wish to extend a sincere thanks to all the members of the planning committee for their contributions in scoping, developing, and carrying out this project.

This report has been reviewed in draft form by individuals chosen for their diverse perspectives and technical expertise, in accordance with procedures approved by the National Academies of Sciences, Engineering, and Medicine's Report Review Committee. The purpose of this independent review is to provide candid and critical comments that will assist the institution in making its published report as sound as possible and to ensure that the report meets institutional standards for quality and objectivity. The review comments and draft manuscript remain confidential to protect the integrity of the process. I wish to thank the following individuals for their review of this report: John Battles, University of California, Berkeley; Patrick Huber, University of California, Davis; Julia Jones, Oregon State University; Lynn Scarlett, The Nature Conservancy; and Ione Taylor, Queens University. Although the reviewers listed above have provided many constructive comments and suggestions, they were not asked to endorse the content of the report, nor did they see the final draft before its release. The review of this report was overseen by Roger Beachy, University of California, Davis. Appointed by the Academies, he was responsible for making certain that an independent examination of this report was carried out in accordance with institutional procedures and that all review comments were carefully considered. Responsibility for the final content of this report rests entirely with the rapporteur and the institution.

Dominic A. Brose
Rapporteur

Contents

1 INTRODUCTION 1

2 DECISION-ORIENTED APPROACHES TO NATURAL RESOURCE MANAGEMENT 7

3 IDENTIFYING NEEDS AND CHALLENGES FOR LANDSCAPE AND MULTI-RESOURCE ANALYSES 15

4 METHODS FOR SPATIAL ANALYSIS: IDENTIFYING SCENARIOS 23

5 METHODS FOR EVALUATING SCENARIOS: RECONCILING QUANTITIES AND VALUES 35

6 MULTI-DISCIPLINARY AND CROSS-AGENCY SYNTHESIS 47

APPENDIXES

A WORKSHOP AGENDA 57

B WORKSHOP PARTICIPANTS 61

C BIOGRAPHIES OF PLANNING COMMITTEE, SPEAKERS, AND STAFF 65

1

Introduction

The responsible management of natural resources for present-day needs and future generations requires integrated approaches that are place-based, embrace systems thinking, and incorporate the social, economic, and environmental considerations of sustainability.[1] Landscape-scale analysis takes this holistic view by focusing on the spatial scales most appropriate for the resource types and values being managed. Landscape-scale analysis involves assessing landscape features in relation to a group of influencing factors such as land use change, hydrologic changes or other disturbances, topography, and historical vegetation conditions. A landscape can be defined as "a large area encompassing an interacting mosaic of ecosystems and human systems that is characterized by a set of common management concerns. The landscape is not defined by the size of the area, but rather by the interacting elements that are meaningful to the management objectives."[2]

As such, different types of data and multiple disciplines may be required for landscape analysis, depending on the question of interest and scale of analysis. Multi-resource analysis (MRA) is an approach to landscape-scale analysis that integrates information among multiple natural resources, including ecosystem services, and is designed to evaluate impacts and tradeoffs between development and conservation at landscape scales to inform public resource managers. This approach implicitly addresses social, economic, and ecological functional

---

[1] A short background paper on landscape-scale and multi-resource analyses was provided to workshop participants and is incorporated as the following Introduction.

[2] Department of the Interior (DOI). 2014. A Strategy for Improving the Mitigation Policies and Practices of The Department of the Interior. A Report to The Secretary of the Interior From The Energy and Climate Change Task Force. [Available at: http://www.doi.gov/news/upload/Mitigation-Report-to-the-Secretary_FINAL_04_08_14.pdf].

relationships; for example, actions to realize the benefits of one type of natural resource (e.g., minerals, oil, and gas) may influence behavior and potential benefits related to other types of natural resources (e.g., recreational opportunities).

The U.S. Geological Survey (USGS) and National Oceanic and Atmospheric Administration (NOAA) are just two of several federal agencies embracing these approaches to encourage broader interdisciplinary thinking across and within their mission areas. The USGS Science and Decisions Center (SDC) is working with partners to develop a framework and proof-of-concept for a next-generation MRA. The MRA builds on USGS resource assessments conducted throughout its long history that provide robust scientific data for land management, water allocation, energy and mineral policy, and conservation of U.S. natural resources. The current suite of resource assessments is conducted under single-discipline assumptions and not readily integrated. Using MRA complements traditional resource assessments by providing an enhanced description of resources-in-place by assessing the impact of extracting natural resources on the depletion or preservation of other natural resources within a defined geographic area. The USGS identifies three major components to a multi-resource analysis (Figure 1-1):

(1) Baseline: integrated information on the current status of multiple natural resources including ecosystem services
(2) Functional Relationships: models describing the interrelationships among collocated resources
(3) Scenarios: analyses evaluating the impacts and tradeoffs to the natural resources in biophysical and socioeconomic terms

The first component is similar to a traditional resource assessment; that is, an inventory of undiscovered and technically recoverable resources. This inventory is enhanced to provide information on multiple natural resources such as oil, water, and pollination in an integrated format. This component includes characteristics similar to a traditional resource assessment such as quantity, quality, and values (or prices) associated with the individual resources; nonmarket values for ecosystem services may need to be considered in this enhanced approach. The second component provides information on the interrelationships among resources in biophysical terms when a resource is developed or extracted. This component moves beyond traditional resource assessments and includes integrated models incorporating the geologic, hydrologic, biological, and ecological sciences. The third component is designed to inform decision makers on the potential biophysical and socioeconomic impacts to the natural resources being studied, given alternative scenarios.

The multi-resource analysis approach responds to guidance described in the Department of the Interior's Secretarial Order 3330[3]; however, SDC has identi-

[3] Secretary of the Interior Sally Jewell issued Secretarial Order Number 3330 "Improving Mitigation Policies and Practices of the Department of the Interior" in October 2013.

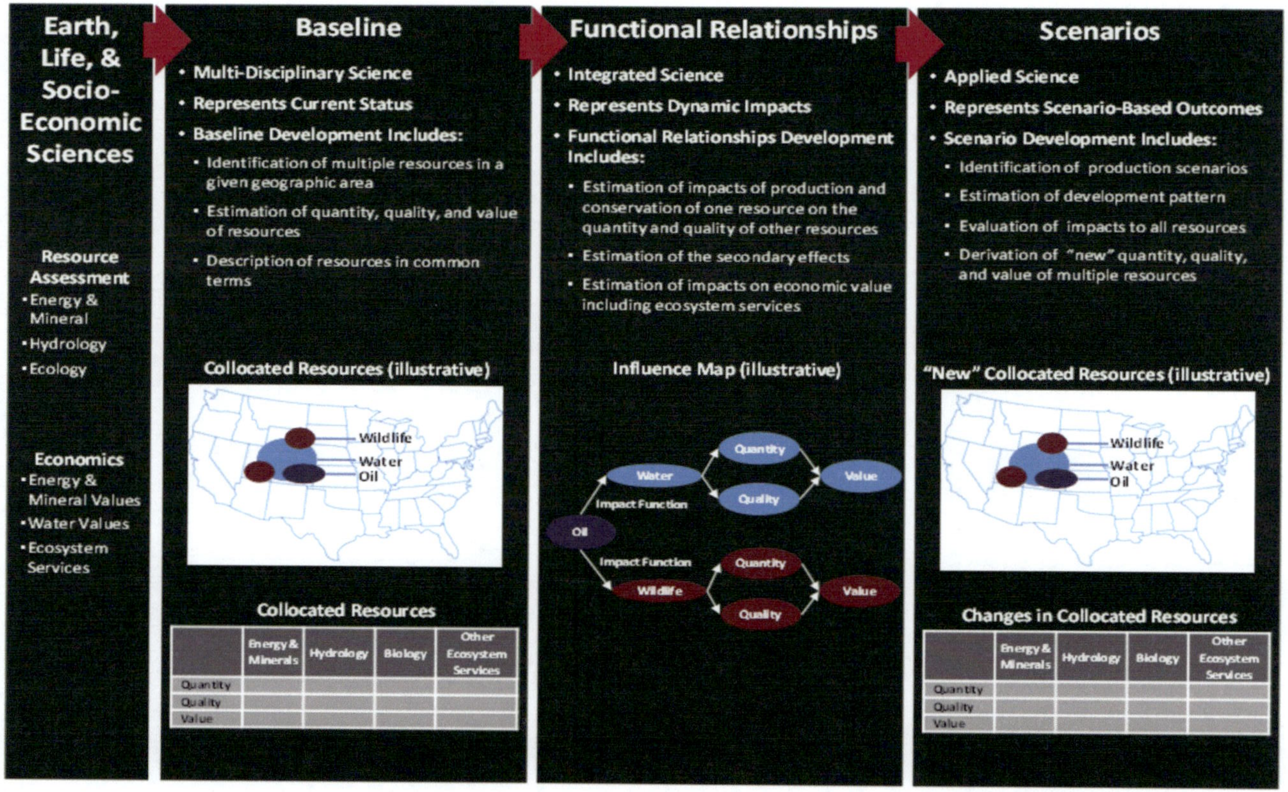

FIGURE 1-1 The three components of multi-resource analysis as identified by the U.S. Geological Survey (USGS).
SOURCE: Image provided courtesy of the USGS.

fied several significant challenges associated with establishing MRA, which include developing

- models or impact functions describing the biophysical consequences of development or extraction of a natural resource on other natural resources and ecosystem services;
- production and recovery functions describing changes in ecosystem service delivery as natural capital is degraded, preserved, or restored;
- economic and management scenarios that are relevant to future decisions;
- studies to estimate changes in monetary and non-monetary values, given scenario-based impacts to multiple natural resources; and
- multidisciplinary teams that communicate and collaborate effectively.

The SDC has initiated two exploratory proof-of-concept projects to advance MRA and demonstrate its feasibility (see Chapter 5). The first project is being developed in the Powder River Basin that is taking a staged approach to develop products to address MRA components. An illustrative model of an integrated, map-based baseline inventory of multiple natural resources has been completed. A series of meetings was held with diverse stakeholders at USGS headquarters and field offices at other agencies including the Bureau of Land Management, U.S. Forest Service, and U.S. Fish and Wildlife Service. The baseline product and conceptual models for the analysis of the interrelationships between natural resources were discussed. Feedback from these meetings is being used to design and develop a pilot-scale model for the Powder River Basin, which will include an integrated analysis of the interrelationships among multiple natural resources under different scenarios.

The second proof-of-concept project the USGS is developing is the Net Resource Assessment (NetRA), which is a scenario-based decision support tool that will enable simultaneous consideration of multiple natural resources to inform national and regional decisions. The NetRA tool will model interrelationships among the natural resources incorporating the economic implications of alternative scenarios. A conceptual framework was developed in 2014 and current efforts are focused on developing a proof-of-concept systems dynamics tool in the Piceance Basin in northwestern Colorado. The tool will estimate the impacts of energy and mineral development on water and biological resources and their ability to provide ecosystem services. The NetRA will enable consideration of multiple natural resources under baseline conditions, examination of their interrelationships, and examination of the impacts if one or more of the natural resources are developed. The NetRA is being developed collaboratively by the USGS, University of New Mexico, Illinois Wesleyan University, and Department of Energy's Sandia National Laboratory.

Building on these examples, the Science and Technology for Sustainability Program of the National Academies of Sciences, Engineering, and Medicine convened a planning committee to organize and hold a workshop on using landscape-based approaches and MRA to better inform federal decision making for the sustainable management of natural resources. The workshop was held on June 2, 2015, in Washington, D.C. The workshop was organized around discussions of knowledge gaps and priority areas for research and presentations of case studies of approaches that have been used to effectively integrate landscape-based approaches and MRA into practice. The workshop was organized into five main panels (see Appendix A: Workshop Agenda):

I. Keynote Panel: Decision-Oriented Approaches to Natural Resource Management
II. Identifying Needs and Challenges for Landscape and Multi-Resource Analyses
III. Methods for Spatial Analysis: Identifying Scenarios
IV. Methods for Evaluating Scenarios: Reconciling Quantities and Values
V. Multi-Disciplinary and Cross-Agency Synthesis

Although the workshop was divided into sequential panels to facilitate discussion of these topics, it should be noted that the panel topics are interrelated and do not follow such an order in practice; USGS recognizes that MRA requires the integration of these concepts. The last panel addressing multi-disciplinary and cross-agency integration, for example, was a discussion of key considerations that need to be addressed at the outset of any MRA. The workshop included participants from federal agencies, policy makers, and the broader scientific community in an exchange of how to incorporate landscape approaches and MRA into the sustainable management of natural resources.

2

Decision-Oriented Approaches to Natural Resource Management

A keynote panel opened the workshop, which included Suzette Kimball, acting director, U.S. Geological Survey (USGS); Steven Ellis, deputy director, Bureau of Land Management (BLM); and Ann Bartuska, deputy undersecretary for research, education, and economics, U.S. Department of Agriculture (USDA). The panel provided a multi-agency perspective on the need for decision-oriented approaches to natural resource management. Dr. Kimball stated the importance of addressing multiple facets of natural resource management and applying decision-oriented science, which is a strategic direction for the USGS looking forward. The multiple facets of natural resource management require working across many disciplines and understanding the potential social and economic impacts to humans. Dr. Kimball said that understanding how to incorporate the human dimension is a central concept to using a landscape approach, and requires recognition that the human dimension is part of the biological underpinning of natural systems.

Department of the Interior Secretarial Order Number 3330 addresses and highlights the importance of applying a landscape-scale approach in evaluating alternative decisions affecting the management and conservation of the nation's natural resources.[1] The USGS has long developed resource assessments of energy and mineral resources and monitored water and biological resources. Decision tools developed from these efforts have benefited federal, state, and other resource managers, decision makers, and policy makers. As natural resources become more constrained, decision makers increasingly demand more sophisticated products

[1] Secretary of the Interior Sally Jewell issued Secretarial Order Number 3330 "Improving Mitigation Policies and Practices of the Department of the Interior" in October 2013.

that describe interrelationships among multiple resources, consider ecosystem services, and evaluate outcomes in biophysical and socioeconomic terms. To address these needs, the USGS is developing multi-resource analysis (MRA), a next-generation product intended to add to, but not replace, existing resource assessments.

Multi-resource analysis has three key components. First, it integrates baseline information across multiple natural resources, which can incorporate the consideration of impacts and tradeoffs at landscape scales for a suite of natural resources, including the potential socioeconomic impacts. Secondly, it explicitly considers change, such as how climate change, urbanization, or extraction of a natural resource may affect other multiple natural resource conditions. It can examine the impacts of a natural resource from the disturbance of another resource in biophysical and socioeconomic terms. Thirdly, specific scenarios are developed and evaluated, which help to examine how multiple natural resources are affected by different decisions. Dr. Kimball stated that the USGS is moving forward with MRA as a new concept that builds on ongoing integrated scientific work and advances decision making. This Academies' workshop, she said, is timely and important because MRA is still in a proof-of-concept development phase and the workshop helps ensure that a community of resource managers, decision makers, and science providers are moving forward in an appropriate way.

The USGS is developing MRA through collaborations with universities, other agencies, and multiple stakeholders in order to gain a better understanding of the nation's natural resources and how the benefits of these natural resources will be affected by change, whether by natural events or by political, societal, and managerial decisions. Dr. Kimball framed these challenges in the form of questions that USGS and others are addressing:

- How can change be effectively considered and incorporated into decision-related products? How are the impacts of land use and climate change, urbanization, and resource extraction across multiple natural resources expressed?
- How can the interrelationships across multiple natural resources be described? Can a functional relationship be developed that quantifies the impacts to multiple natural resources from a disturbance to one of those resources?
- How can the consequences to humans, along with biophysical consequences, be more fully considered? To what extent can diverse impacts across multiple natural resources be valued so that comparisons and tradeoffs are considered?
- How should scenarios be identified and what types of scenarios are most useful to decision makers? To what extent should future resource assessment products include scenario analysis?

Mr. Ellis shared his perspective from 35 years of government service. As a forest supervisor for the Freemont-Winema and Wallowa-Whitman National Forests, Mr. Ellis was a decision maker with direct experience writing and executing land use plans. The BLM, Mr. Ellis explained, is an agency that is 200 years old with 245 million acres of public land and until the Federal Land Policy and Management Act of 1976 (FLPMA), the BLM did not engage in land-use planning. The goal until then, after the national parks and forests were delineated, was primarily to dispose of the lands to private owners for development. With the introduction of FLPMA, land-use plans were developed, but not by using a landscape-based approach. Generally, he said, they followed the administrative boundary of a park or forest.

The scope of early land use planning primarily focused on allocating resources, determining right of ways for power lines, determining locations for livestock grazing, and managing timber. These early land-use plans did not have ecological or social elements. In time, the BLM issued more comprehensive land-use plans, engaged with the public, and allowed more public input. Mr. Ellis said the BLM now engages the public collaboratively upfront during development of a land use plan. The BLM has also begun focusing on resource and science-based decisions to align land-use planning with multiple-use goals and sustained yield, which is the use of resources for the benefit of all in a manner that would not impair the resources' future productivity. Land-use planning was no longer within one administrative unit at the BLM, but was taking on a landscape-based approach.

Mr. Ellis said the need for moving across political and administrative boundaries was driven by the recognition that resources, such as wildlife, crossed such boundaries and needed to be managed more holistically. For example, the BLM, in cooperation with the U.S. Forest Service, recently developed a series of environmental impact statements (EIS) to incorporate greater sage-grouse conservation measures into BLM land-use plans. These EISs focus on priority habitat with the highest value to maintaining sage-grouse habitat. Moving forward, the BLM is implementing Planning 2.0, which increases public participation in the development of BLM's Resource Management Plans and seeks to move beyond jurisdictional boundaries to better manage resources. The BLM is also seeking collaborations with other agencies to better incorporate science into land-use planning. The BLM does not have a science branch, and looks to the USGS and Fish and Wildlife Service to provide the scientific foundations needed for improved decision making.

Dr. Bartuska offered that in developing and implementing land-use plans, the scale of analysis must be commensurate with the scale of the decision. It is important to connect an analysis to the practical dynamics of a decision, such as regulatory or legal considerations. The USDA, she said, has moved to 21st century conservation, which addresses conservation at the landscape scale by entering into partnerships and using actionable science. The USDA is continually developing innovative solutions to conservation across public and private lands. The Natural Resources Conservation Service (NRCS) Landscape Conservation Initiatives are a great example of such efforts, she said. The Landscape Conservation Initiatives were established under the 2008 Farm Bill, and aim to accelerate results from voluntary conservation programs. The programs are primarily driven by grassroots input and local needs, and seek to enhance locally driven processes to better address nationally important conservation goals.

The Mississippi River Basin Healthy Watershed Initiative, Dr. Bartuska said, is an example of an innovative solution to conservation. The program is implemented through the NRCS and is designed to work with producers and landowners to implement voluntary conservation practices that improve water quality, restore wetlands, enhance wildlife habitat, and sustain agricultural profitability in the Mississippi River Basin; however, these goals are accomplished by focusing not on the whole basin but on targeted watersheds within the larger Basin. Dr. Bartuska said that improving management practices in targeted watersheds has a large influence over the overall health of the Mississippi River Basin.

Monitoring to assess progress on improving water quality occurs at three scales. Water quality monitoring at the edge of fields assesses how practices influence nutrient loading at the local scale, monitoring in the main stream of the Mississippi assesses the impact on water quality on a large scale and essentially aggregates results from across many fields, and monitoring across time can provide the long-term trajectory of water quality improvements in the Mississippi River Basin. Dr. Bartuska highlighted the importance of the scale of the assessment, and assessing where in practice the decisions are made and at what level are key aspects to a successful program. It is also important to include the social context in the decision-making process. In 2010, there were approximately 800 watershed councils within the Mississippi River Basin, which are private entities composed of stakeholders that function as the governance system. Any conservation program or management practice being implemented at the watershed level will need buy-in from these stakeholders.

Part of the challenge of using any landscape- or multiple resource analysis-based approach is the complexity of landscapes that need to be integrated into a holistic picture. An actively managed landscape, such as an agricultural region or forest, will have elements of natural landscapes embedded but will also have societal influences acting on the decision-making process. It is critical for the decision to be science-informed and for an analysis to cut across many scales while acquiring the data necessary to support meaningful decision making, Dr. Bartuska said. The Agriculture Research Service has developed the Long-Term Agroecosystem Research (LTAR) Network, which links university partners to approximately 23 benchmark experimental watersheds that collect long-term data on agricultural sustainability, climate change, ecosystem services, and natural resource conservation at the landscape scale. The LTAR Network allows for more extensive analysis and utilization of a vast collection of data, and will help guide decisions that are needed on the management of natural resources.

As Dr. Bartuska explained, the U.S. Forest Service developed the Watershed Condition Framework Process, which establishes a new process for improving the health of watersheds on national forests and grasslands. The Watershed Condition Framework Process facilitates new investments in watershed restoration in a way that intends to provide more economic and environmental benefits to local communities. The process is akin to adaptive management at the watershed scale, and is transferable across agencies and organizations (Figure 2-1). Dr. Bartuska explained that Step D in Figure 2-1 (Implement Integrated Projects) helps provide consistency across planning units and creates the basis for developing projects. Steps E and F, she explained, are the most important because they address tracking and reporting on restoration accomplishments through monitoring. These

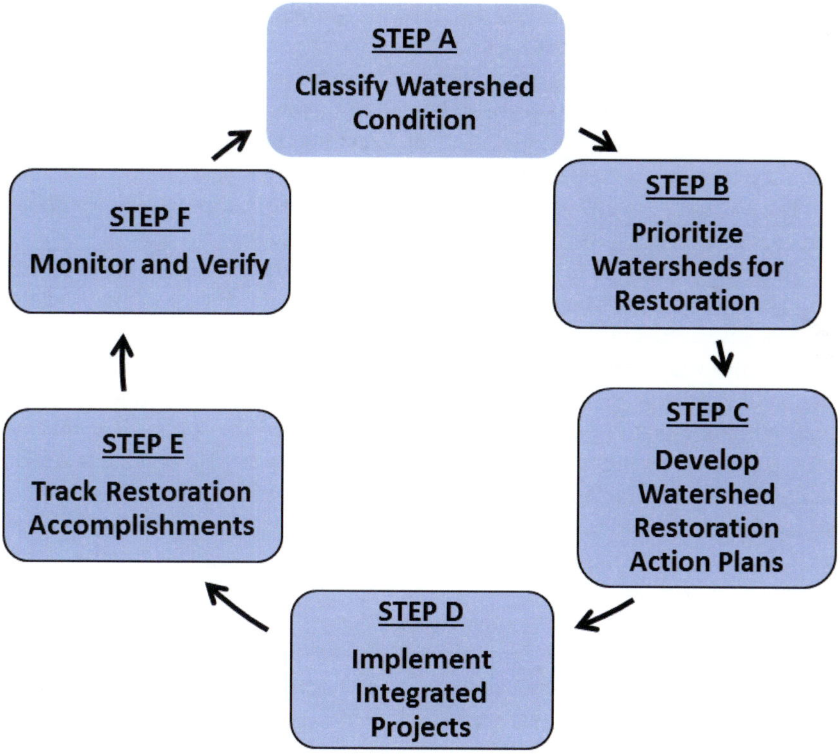

FIGURE 2-1 Watershed Condition Framework Process.
SOURCE: Ann Bartuska, Presentation, National Academies of Sciences, Engineering, and Medicine Workshop, June 2, 2015, Washington, D.C.

steps provide the feedback necessary to adapt for the next cycle of developing watershed goals.

During the question-and-answer period, the panel was asked about how federal agencies manage collaborations that engage the local public where a decision may be implemented while also being responsive to national stakeholders. Mr. Ellis responded that it depends on the scale of the collaboration, and that most importantly the collaboration must have a clear mission. Collaborations are most successful when all parties agree on the goal from the outset. The BLM, for example, was successful in implementing travel management plans in eastern Oregon that focused on determining appropriate locations for the use of off-highway vehicles (OHV) by engaging the public locally and being clear that the mission was to protect the areas where OHV use was not appropriate.

Another question related to looking across academic studies for long-term research as a means of increasing replication for a given study and gaining greater insight into the data. Dr. Bartuska stated that two research sites in the LTAR Network are not federal facilities but are academic institutions. A challenge to long-term research, she added, is the difficulty in maintaining research for an extended period of time outside of the federal sector. One of the National Science Foundation-funded long-term ecological research sites (LTER), for example, has been in operation for 30 years and faces an ongoing challenge in maintaining and keeping the site operational. A different approach to meeting the same goal of cutting across academic and federal sectors to gain insight into available data is by keeping long-term sites in the federal system, but providing access to academic scientists.

Another dynamic of long-term research, Dr. Bartuska added, is the need for participatory research, where the public is brought into the design and implementation of research, and the monitoring and analysis of results. There are useful opportunities when there are controversial issues to expand the engagement to the public rather than keeping it in the scientific space, and long-term research sites could allow for that opportunity. Paul Sandifer echoed the support for greater participation of the public into scientific research projects. He commented that there has been a political backlash against supporting social science in decision making, and that there instead should be expanding support for the social sciences that in turn supports science-based decision-making processes that directly affect natural resource management.

The panelists were asked to describe challenges with interagency collaborations and in engaging stakeholders, and also about the need to strengthen capacity within agencies to engage stakeholders on a sustained basis. Dr. Bartuska responded that within USDA, there are Forest Service-related committees but also agriculture-related committees, which creates an internal division that separates regulatory- and appropriation-related activities. It is unlikely for that structure to change, so the challenge is in bridging those internal divisions in order to execute effective collaborations. Breaking down institutional barriers, Dr. Bartuska said, is the largest challenge; however, eliminating those barriers so that personnel can work across agency boundaries to share resources and budgets would have a significant impact on more effectively executing interagency collaborations. She added that agencies should also have stakeholder engagement programs, and highlighted the Resource Conservation and Development (RC&D) program within the NRCS as an example. The RC&D program funded local volunteer councils that helped communities to protect and develop their economic, natural, and social resources while also improving their area's economy, environment, and quality of life. This program focused mostly on private lands, and focused an NRCS employee on helping to write grants that would build a network to meet the needs of the local community. The program was successful because it capitalized on local leadership and knowledge; however, they were unfunded nearly three years ago. Since then, she added, successful partnership-based programs have emerged in various forms that seemingly use the same model. It is the local leader who, when supplied with the right resources, can bring together a successful network.

Mr. Ellis said that he has not had a challenge working across agencies when there are enough similarities between mission areas. For example, he described the Forest Service and BLM as having similar structure, legal mandates, and land they manage. Mr. Ellis said that the best collaboration he experienced was in Wallowa County, Oregon, which was a collaboration that included the general public, the Nez Perce Tribe, Oregon State University Extension, county government agencies, and the livestock industry. Mr. Ellis said that in over six years of being involved with the collaboration, there was never an issue they were not able to work through because of the leaders involved in the collaboration. They were dedicated to the collaboration and were able to set aside personal agendas in order to focus and achieve common goals. An agency's capacity for stakeholder engagement, he added, often depends on the timeframe. The BLM implemented plans under court-ordered timelines, which were too restrictive to appropriately engage stakeholders and made the implementation of the plans challenging.

Dr. Kimball commented that a challenge to collaborations is due to the many layers of an agency. Field scientists, she said, can have relationships with community and local groups that work very well, but as one moves up the layers of an agency, systemic barriers are added so that at the national headquarters level there are budgetary and political barriers that can disrupt those functional relationships. She highlighted a previous experience when USGS and the National Oceanic and Atmospheric Administration (NOAA) entered into a memorandum of understanding; however, it took three years to transfer funds from one agency to another. The scientists at the local level had abandoned the collaboration by finding ways to work around the agreement in order to accomplish the work they had set out to do. The budget process, she said, is often the single most important barrier to collaborations. Several agencies may be interested in collaborating on a project and each will put in their respective budget requests, but only one of the agencies will receive the funding requested. Often different agencies are reviewed by different Office of Management and Budget (OMB) examiners, therefore requiring that different avenues be taken in order to receive the necessary funding. Dr. Kimball said this adds levels of complexity and delays in implementing multi-agency collaborations.

3

Identifying Needs and Challenges for Landscape and Multi-Resource Analyses

Kit Muller, management and program analyst at the Bureau of Land Management (BLM), provided five characteristics necessary to fostering successful landscape-based collaborations:

- Shared goals
- Shared understanding of risks and potential tradeoffs
- Shared commitment to the undertaking required for management actions
- Shared commitment to evaluating and reporting on progress
- Shared commitment to adapting to new information

The shared understanding of risks and potential tradeoffs, he added, is a key characteristic of successful multi-resource analysis (MRA). Managing a collaboration is often more challenging than the scientific research—the social context is important.

Mr. Muller discussed a series of proposed Sage-Grouse and Sagebrush Conservation Plans as a new way for the BLM to operate. Sage grouse across the western states have declined in number over the past century due to the loss of sagebrush habitats. Greater sage-grouse habitat covers 165 million acres across 11 states, which represents a 56 percent loss in the historic range of sage-grouse habitat; the BLM estimates that the sage-grouse population numbered in the millions but declined to between 200,000 and 500,000. The BLM, U.S. Forest Service, Fish and Wildlife Service, and Natural Resources Conservation Service (NRCS) are developing the proposed Sage-Grouse and Sagebrush Conservation Plans to ensure the protection of remaining habitat. Planned actions focus on three major management strategies: manage disturbance, restore habitat, and reduce the risk of wild fire. These plans are also a framework for conservation and development in the sagebrush biome on which further efforts can be built upon.

One example of the BLM building on such a framework is the development of a mobile-app that helps to track disturbances online at a project level and helps land managers better assess intactness across the landscape. Mr. Muller said that if the BLM is going to only allow a limited amount of disturbance per tract of land, then the agency will have to either reduce an overall amount of disturbance or stop authorizing new disturbances. Also building on the framework approach, the BLM and USDA have committed to better monitoring and reporting by agreeing on a core set of indicators and standard methodologies for collecting data pertaining to terrestrial conditions.[1] By combining sampling efforts, the BLM and NRCS can report on a wide range of terrestrial conditions. Another opportunity, Mr. Muller said, would be to build on existing assessments conducted in support of the Sage-Grouse and Sagebrush Conservation Plans in the Great Basin. A larger research effort could be developed around identifying ecosystem services provided by the implementation of the plans and identifying gaps in ecosystem services that remain. A similar analysis of existing assessments would also help to identify other gaps in the scientific knowledge pertaining to conservation, and further the use of the plans as a framework for building conservation and development efforts in a large biome.

Elsa Haubold, National Landscape Conservation Cooperatives (LCC) coordinator for the U.S. Fish and Wildlife Service, presented several examples from the LCCs to provide a perspective on challenges to implementing multiresource analysis (MRA).[2] Twenty-two LCCs were created by the Department of the Interior in 2009 with Secretarial Order 3289.[3] The LCCs form a network of resource managers and scientists that work across jurisdictions to better integrate science and management into conservation. Each LCC is a self-directed partnership of federal, state, and local governments along with tribes and first nations, nongovernmental organizations, universities, and interested public and private organizations that work collaboratively to identify best practices, connect efforts, identify science gaps, and avoid duplication through conservation planning and design. Although the Fish and Wildlife Service leads a majority of the LCCs, there is one in the Caribbean led by the U.S. Forest Service and others co-led by the Bureau of Reclamation and National Park Service. Dr. Haubold explained that MRA can be defined and integrated into an overarching framework to guide deci-

[1] MacKinnon, W.C., J.W. Karl, G.R. Toevs, J.J. Taylor, M. Karl, C.S. Spurrier, and J.E. Herrick. 2011. BLM core terrestrial indicators and methods. Tech Note 440. U.S. Department of the Interior, Bureau of Land Management, National Operations Center, Denver, CO [Available online: www.blm.gov/nstc/library/pdf/TN440.pdf].

[2] Available at: https://nccwsc.usgs.gov/content/landscape-conservation-cooperatives-lccs.

[3] Former Secretary of the Interior Ken Salazar issued Secretarial Order Number 3289 "Addressing the Impacts of Climate Change on America's Water, Land, and Other Natural and Cultural Resources" in September 2009.

sion making. The phrase "guide decision making," she said, is critical and is an issue the LCCs are trying to better address. By incorporating decision makers early in a process, a decision-making guidance product can be developed that is clear in its conservation objective and in how the decision maker should use the product.

The South Atlantic LCC spans several state boundaries from Virginia to Florida and developed a mission to create a shared blueprint for landscape conservation actions that sustain natural and cultural resources. The LCC developed 29 ecosystem indicators and targets, which included targets such as improved habitat and individual species' populations.[4] The LCC held regional blueprint workshops to identify priority landscapes and integrate existing conservation plans. Dr. Haubold emphasized that there is a tremendous amount of existing information available that needs to be integrated into the planning process. The LCC launched the blueprint in 2014 and already issued an updated version for review. To make the planning process more quantitative, the LCC updated the revised blueprint with more data.

Dr. Haubold said that the key relationships among resources that need to be incorporated into MRA depend on the objectives for and scale of the landscape. She provided a case study of the Upper Midwest and Great Lakes LCC, which is restoring the connectivity between the Great Lakes and their tributaries. The challenge to meeting this objective is the over 270,000 barriers, such as roads and dams; however, many of these tributary barriers are necessary to prevent the spread of invasive species, such as the Asian carp. The Upper Midwest and Great Lakes LCC determined which barriers provide the most benefit and which ones could be removed. A model was developed that optimized the amount of habitat gained with dollars spent based on barriers identified by the LCC. Incorporating the Department of Transportation into the assessment and decision making was critical because of their mission to oversee and maintain roadways.

Datasets needed for MRA, similar to the necessary key relationships, also depend on the scale of the objectives to be attained. The South Atlantic LCC, for example, needed current land cover to attain their objective of developing a shared blueprint for landscape conservation actions. Dr. Haubold described the Appalachian LCC's priority of assessing energy development in the Appalachian region. The Nature Conservancy (TNC) developed the Pennsylvania Energy Impacts Assessment for the LCC that assessed the potential for energy development in the region;[5] however, TNC did not start from scratch with datasets, she said. There were existing datasets and information made available by establishing a rich collaborative of stakeholders and different communities.

TNC's assessment identified areas with a high probability of having over 50 years of wind, shale gas, and surface coal mining development as they pertain to intact forests (Figure 3-1). Similarly, the assessment identified energy develop-

[4] Available at: http://www.southatlanticlcc.org/indicators.

[5] Available at: www.nature.org/media/pa/tnc_energy_analysis.pdf.

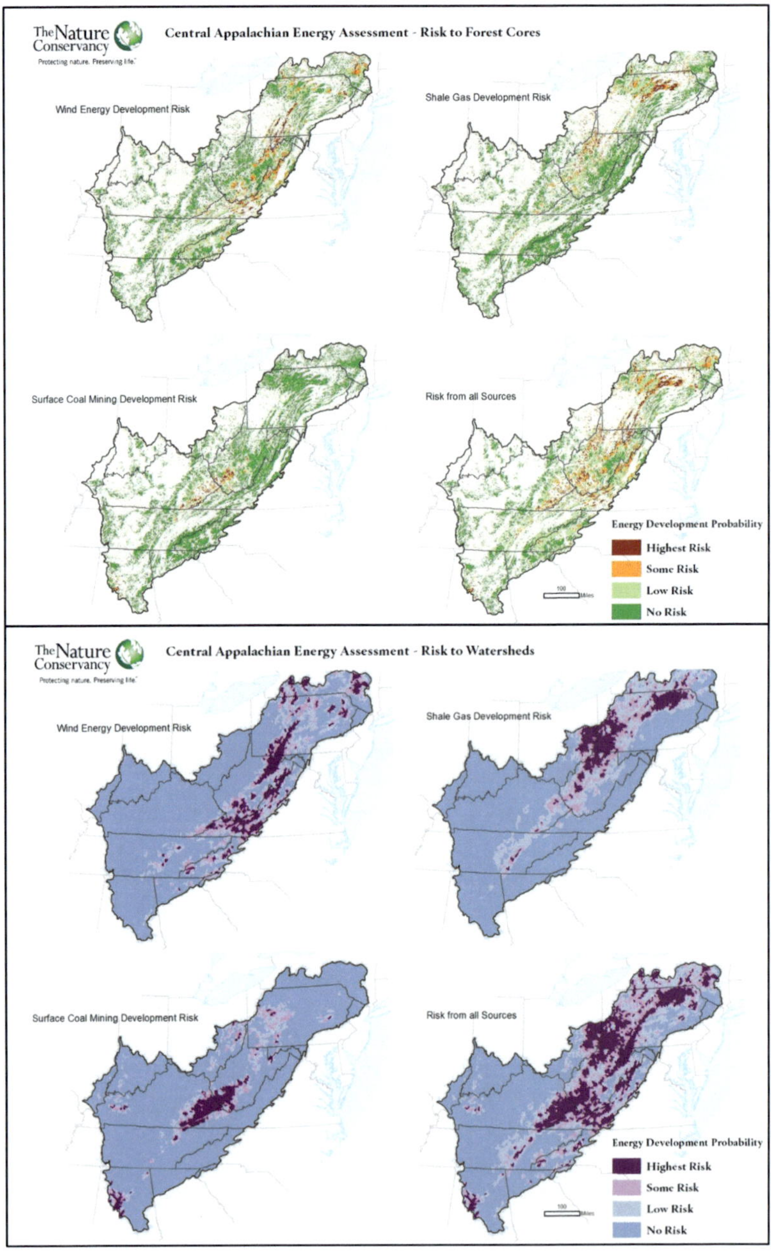

FIGURE 3-1 Modeling outputs for intact forest and energy develop probability (top) and for watersheds and energy development probability (bottom).
SOURCE: Elsa Haubold, Presentation, National Academies of Sciences, Engineering, and Medicine Workshop, June 2, 2015, Washington, D.C.

ment as it pertains to watersheds. The assessment also identified watersheds of high importance for drinking water supply, and watersheds of high importance with the scenario of the highest level of energy development. This analysis was critical for communities to assess how energy development could potentially impact drinking water supplies.

Dr. Haubold summarized by emphasizing that no single organization can tackle future challenges. A landscape-based framework to guide decision making requires bringing partners together to develop shared visions, goals, and objectives; identify existing datasets and data gaps; and achieve objectives collectively by sharing resources and overcoming jurisdictional barriers. Key relationships among resources depend on the scale of the landscape, and need to be identified by the partners in the collaboration. Datasets should have some level of standardization in data collection methodology and the type of data being collected. This would allow for data across a landscape or multiple landscapes to be more easily and accurately compared. Stakeholder fatigue is another key consideration when establishing collaborations due to limited staff capacity within most agencies. Finally, landscape conservation, she said, occurs locally, but must be thought of regionally.

Ione Taylor, executive director of Earth and Energy Resources Leadership at Queens University, discussed challenges to and new approaches for MRA. Natural resource modeling, she said, is changing from descriptive, monitoring-based approaches to multi-discipline and multi-purpose assessments that integrate information across scenarios (Figure 3-2). Similarly, model layering is incorporating more interactivity and interoperability of data and other models; however, Dr. Taylor noted that progress in the interconnectivity of models occurred in a patchwork way and not as a whole system.

Dr. Taylor described integrating MRA as a disruptive technology with many considerations, including the following:

- Consideration of both monetary and non-monetary value of all products, services, and processes (e.g., natural capital, ecosystem services)
- Multiple, inter-connected resources and their associated cycles
- A wide range of disciplines (i.e., geoscience, engineering, business, economic, environmental, stakeholder, policy, and regulatory)
- Goals to improve risk assessment and decision-making ability despite uncertainty and complexity
- Recognition that risks are involved and rewards are uncertain

The U.S. Geological Survey (USGS), for example, seeks to integrate hydrology, geology, economics, biology, and ecosystems into derivative products and services that take a higher-level approach to informing decision making. This requires overcoming the polarization that occurs between extractive industries interested in developing natural resources and those interested in conservation.

Trends in Natural Resource Models

Moving from......	Going toward......
Descriptive monitoring and reporting on individual elements of natural systems	Multidiscipline, multipurpose assessments of natural systems; developing response strategies
Descriptions of state	Forecasts and scenario development
A few first-generation multifunctional multidisciplinary products and services	Routine way of doing business
Systematic 2 D mapping	Dynamic 3 D models
Layering of models	Interactivity and interoperability of data and models
Grab-bag of products with scale and content gaps	Seamless suite of products spanning all temporal and spatial scales
Ad hoc, stand alone efforts	Multi-agency, multi-stakeholder programs

FIGURE 3-2 Trends in natural resource models.
SOURCE: Ione Taylor, Presentation, National Academies of Sciences, Engineering, and Medicine Workshop, June 2, 2015, Washington, D.C.

The challenge, Dr. Taylor said, is that conservationists view the economy as a wholly owned subsidiary of the environment, while industry views their impacts as environmental externalities. Most of society, however, lies in the middle of this polarization and is composed of consumers that wish to reduce their impact on the environment. MRA can break this polarization and can be considered a disruptive technology.

Another challenge to conducting MRA is the silos that exist within and among federal agencies. Horizontal integration must occur by opening those silos and determining which subsets of individual disciplines need to be incorporated into the process, she said.[6] An analytical decision science framework and model helps to facilitate this horizontal integration by changing the process from one that is provider-driven (i.e., economist or scientist) to a process that is client-

[6] Although not presented by Dr. Taylor, the USGS was a sponsor of the recent NRC report that addresses the issue of horizontal integration across disciplines and federal agencies: National Research Council (NRC). 2013. Sustainability for the Nation: Resource Connection and Governance Linkages. Washington, D.C.: The National Academies Press. [Available: http://www.nap.edu/catalog/13471/sustainability-for-the-nation-resource-connection-and-governance-linkages].

driven (i.e., land or resource manager). A broader framework that integrates across all the silos within or among agencies, Dr. Taylor said, will deliver a more useful product for a land or resource manager.

Dr. Taylor commented that approaching a problem at the right scale is a critical component of defining boundaries around a decision. Most resource work is conducted at a single scale of analysis using data collected at that scale when a multi-scale-based approach is needed. Scaling up is also a challenge because only some data types are additive, and variability and complexity of natural systems may be lost when data that are not additive are aggregated. The scale at which data are collected and aggregated most frequently may not be useful for the question at hand. One of the most difficult tasks is determining what level of complexity is needed and which components need more or less complexity. There is a physical scale to a scenario, she added, but also a temporal scale. Using knowledge derived from outcomes at a small scale to predict and manage large-scale phenomena can be challenging, because outcomes at a large scale are likely to have characteristics that cannot be predicted from the outcomes derived from the small scale. Similarly, outcomes at a small scale are likely to have characteristics that cannot be predicted from outcomes derived at a large scale. Large-scale processes and relationships mask the variability that exists at smaller scales.

At the temporal scale, slow-moving variables are considered external to a system for simplification in order to focus on fast-moving variables when studying short periods of time. Slow moving variables, at times, are not important, but can be key drivers to a system at other times. Dr. Taylor noted that time is an important consideration, and has worked with managers within BLM who were very interested in assessing the impact variables have cumulatively through time as decisions are made.

Dr. Taylor also commented on the social elements necessary to integrate different aspects of MRA in order to develop the most useful product. There is a disconnect between the subject matter expert developing a dataset and the land use manager making decisions based on that data. There is a range in agreement with a decision that is a function of the uncertainty in the decision; the larger the uncertainty, the less agreement there will be with the decision. Often, she remarked, the degree of certainty with a decision depends on the complexity of the issue; however, it is critical for both the subject matter expert and the land use manager to change their mindsets so that all parties can move towards agreement even with great uncertainty. It is expertise, however, that hinders their ability to change their mindsets, but change is critical in order to optimize the process and work towards a successful decision.

The Center for Creative Leadership developed a model called Boundary Spanning Leadership, which guides expert groups on how to span boundaries (Figure 3-3).[7] First, it is necessary to bring the right subject matter experts to-

---

[7] www.ccl.org/Leadership/landing/spanboundaries.aspx.

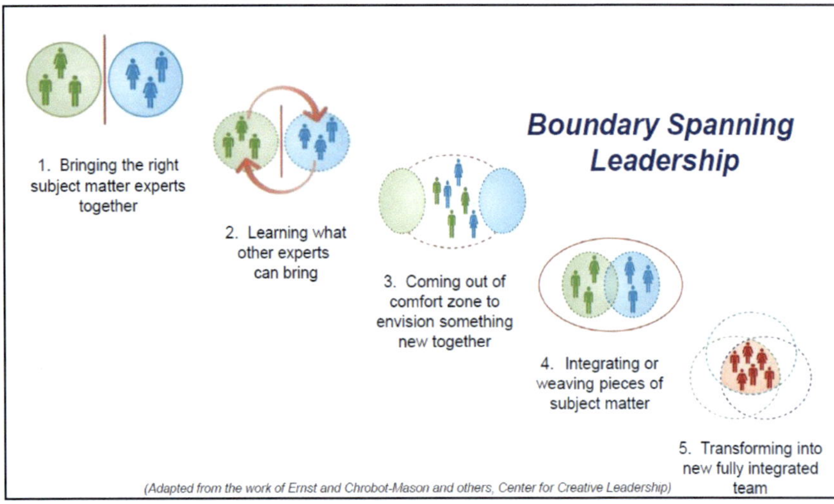

FIGURE 3-3 Boundary Spanning Leadership developed by the Center for Creative Leadership.
SOURCE: Ione Taylor, Presentation, National Academies of Sciences, Engineering, and Medicine Workshop, June 2, 2015, Washington, D.C.

gether. Dr. Taylor noted that this step requires listening to experts and recognizing where there is overlap in fields represented by the group. Next, the experts in the group need to leave their comfort zones to collectively envision something new. The final two steps are to begin weaving the subject matter pieces together and transforming the group's vision into a new, fully integrated team. Dr. Taylor provided an example of the application of Boundary Spanning Leadership during initial work on an MRA from the Powder River Basin. Her expectation with the MRA was that the geographic boundaries of the analysis would be agreed upon by the group quickly, and would be composed of the outline of the mineral deposit in the subsurface as projected onto the land surface. Another participant, however, strongly felt that the boundaries should be the range of habitat of a potentially endangered species. Dr. Taylor said that in facilitating the work of the team she already had moved to the fourth step of the Boundary Spanning Leadership model integrating the pieces before the first few steps of understanding the group dynamics. It took several more hours of discussion among the group before the geographic boundaries of the MRA were decided upon. Dr. Taylor concluded that the Powder River Basin MRA was an important learning experience, and that there needs to be more training for senior management to understand how to break down boundaries to more efficiently integrate concepts and reach the fifth step of the process—transforming into a new, fully integrated group.

4

Methods for Spatial Analysis: Identifying Scenarios

Lisa Wainger, research professor at the University of Maryland, discussed optimization modeling to analyze multi-resource management goals. She identified three elements to developing scenarios: economic efficiency, legal compliance, and social equity. Economic efficiency states that parcels (resources) will be utilized at their highest and best use. In practice, this translates to landowners using their land in a way that provides the highest income over time. Increasingly, however, public agencies are finding opportunities for preserving more land. Government protection creates a market condition to preserve agricultural land. Part of economic efficiency, she said, is the concept of cost-efficient landscapes and the maintenance of networks, energy transmission, and transportation on a large scale. Legal compliance generally relates to natural resource and human health protection, and to limits on incompatible land uses and environmental externalities. Social equity addresses the fair distribution of resources among society. Dr. Wainger presented an optimization model used to generate scenarios (Figure 4-1). The model was originally developed to address the need to meet a total maximum daily load of a pollutant by optimizing where on a landscape to place certain best management practices. The model represents a parcel of land or a landscape with certain constraints. One constraint was keeping producer profits neutral, meaning if something was done that impinged on another's private profits, then there would be less likelihood that the action would be adopted. Other input to the model included available parcels and actions, costs and performance, and ecosystem service goals.

In describing the model, Dr. Wainger said that available parcels and actions address what land is available to convert to other land covers and other actions on the landscape. Those conversions and actions are then characterized in terms of

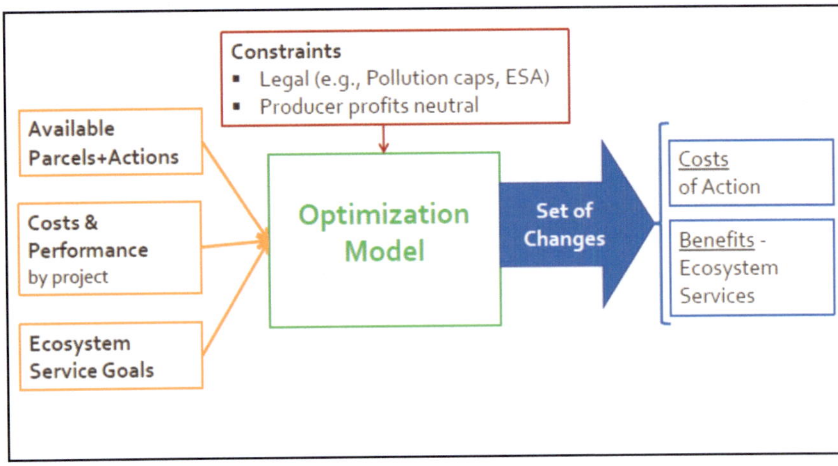

FIGURE 4-1 The components of an optimization model used to generate scenarios. SOURCE: Lisa Wainger, Presentation, National Academies of Sciences, Engineering, and Medicine Workshop, June 2, 2015, Washington, D.C.

their costs and performance. For example, if the scenario was to develop the land, then the cost to convert the parcel and the monetary gain expected would be calculated. If instead it was a best management practice to reduce pollution, then the cost for implementation and the benefits achieved would be calculated. Dr. Wainger said the model was developed with ecosystem goals that were measurable. For example, the co-benefits of reducing pollution separate from the benefits achieved with legal requirements were quantified. Overall, the analysis from an optimization model reflects what the user perceives as important goals to set. The model computes all inputs to provide outputs that best meet all the user's requirements while also reporting the results in terms of costs and benefits. The scenarios can vary by adjusting specific variables or more generally by focusing on broader goals, such as social equity. When using optimization models, she said, dollar values can be used, but other metrics can also be used as indicators of benefit. Such benefit indicators are based on economic principles, and can highlight which characteristics may cause harm or generate benefits.

Dr. Wainger presented case studies that address issues connecting the biophysical, social, and economic considerations as well as cumulative impacts. The first case study focused on cheatgrass in the Twin Falls District of southern Idaho.[1] After cheatgrass areas in western grazing lands burn, mostly due to natu-

[1] Wainger, L. A., King, D. M., Mack, R. N., Price, E. W., & Maslin, T. 2008. Prioritizing invasive species management by optimizing production of ecosystem service benefits. USDA Economic Research Service. [Available online: http://naldc.nal.usda.gov/download/32824/PDF]

ral processes, the BLM and USGS collaborate to restore the sites in order to prevent the intrusion of cheatgrass, which can dominate an area shortly after a fire. Dr. Wainger used an optimization model to analyze how a bundle of ecosystems services are maximized instead of just one service. Because it was not possible to have optimal grazing or sage-grouse habitat, the model was used to analyze how the two could coexist.

Under scenario A, the model found parcels that delivered multiple ecosystem service benefits at high levels, such as habitat protection, grazing, antelope hunting and recreation, and property protection from fire (Figure 4-2). Under scenario B, the model optimized conditions where agencies selected sites viewed as superlative for one service at time. Overall, the model outputs demonstrated that considering multiple benefits at once was more cost-effective. Successful restoration and prevention of cheatgrass was incorporated into the model as well as the actual decision process used by BLM. For example, one site under scenario B was selected because it was important for preventing fires, but did not have other co-benefits. Because the decision process BLM was using was incorporated, Dr. Wainger said, the model served as a decision support tool and not a decision-making tool.

Dr. Wainger presented another case study of an analysis on mining predominately in West Virginia that assessed an extreme scenario of 100 percent of mine permits being utilized and an intermediate scenario of 20 percent of mining permits being utilized (Figure 4-3).[2] The analysis focused on one ecosystem service—freshwater angling benefits. The analysis incorporated biological models for determining game fish effects under mining scenarios and the resulting demand for fishing, which is considered a benefit indicator (Figure 4-4). Using a benefit transfer of fishing effects, monetary damages were generated from the scenarios. Dr. Wainger said the analysis demonstrated that if monetary decisions are based on one benefit, the spatial heterogeneity of the benefit can be used to separate out incompatible uses. Willingness to pay values, however, were not large relative to the value of coal, which shows the importance of trying to not meet all needs from one geographic area. The analysis demonstrated that recreational fishing impacts were fairly low in areas where mining was spatially separated from the population; however, there were clear resource-use conflicts when mining occurred in the more populated parts of the study area with recreational fishing.

Dr. Wainger summarized key concepts learned with optimization modeling. Scenario analysis can demonstrate preferred scenarios given a set of goals and constraints, how to best meet those goals, the co-benefits associated with meet-

[2] Mazzotta, M., L. Wainger, and S. Silfleet. Assessing Lost Ecosystem Service Benefits Due to Mining-Induced Stream Degradation in the Appalachian Region: Economic Approaches to Valuing Recreational Fishing Impacts. Presented at Northeast Ag/Resource Economics Assoc, Morgantown, WV, June 1–3, 2014.

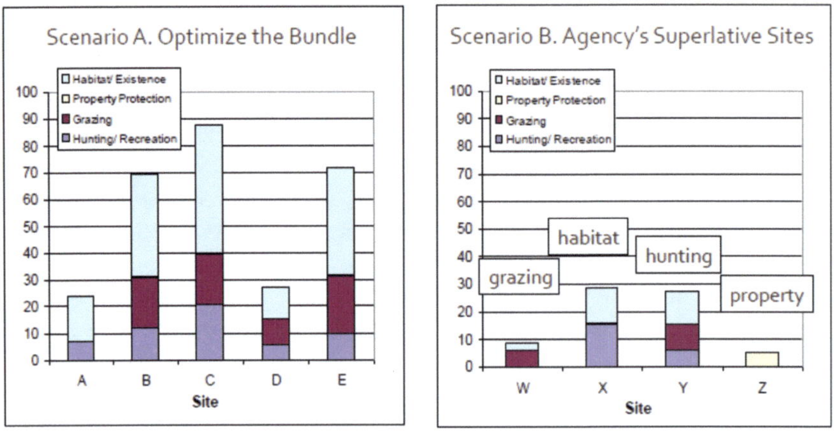

FIGURE 4-2 Output from scenarios A and B of the optimization model for restoration of burned areas to prevent cheatgrass intrusion.
SOURCE: Lisa Wainger, Presentation, National Academies of Sciences, Engineering, and Medicine Workshop, June 2, 2015, Washington, D.C.

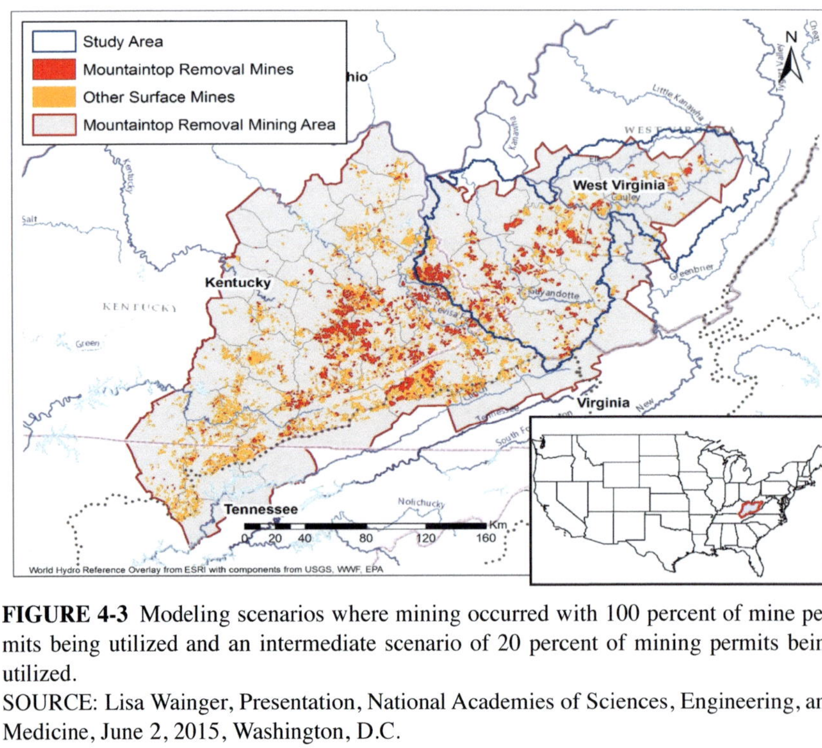

FIGURE 4-3 Modeling scenarios where mining occurred with 100 percent of mine permits being utilized and an intermediate scenario of 20 percent of mining permits being utilized.
SOURCE: Lisa Wainger, Presentation, National Academies of Sciences, Engineering, and Medicine, June 2, 2015, Washington, D.C.

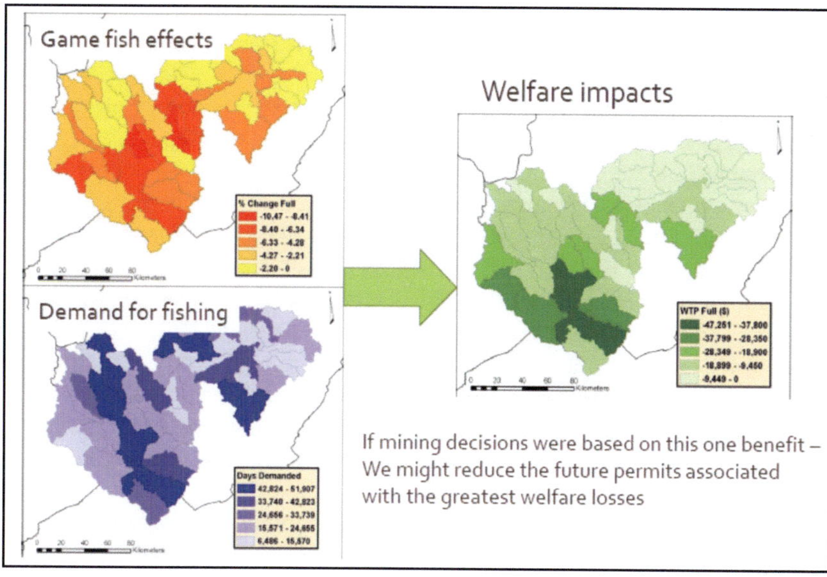

FIGURE 4-4 When monetary decisions are based on one benefit, the spatial heterogeneity of the benefit can be used to separate out incompatible uses.
SOURCE: Lisa Wainger, Presentation, National Academies of Sciences, Engineering, and Medicine Workshop, June 2, 2015, Washington, D.C.

ing those goals, and the costs associated with policy decisions. A landscape-scale analysis, cautioned Dr. Wainger, can be very sensitive to market fluctuations. For example, a seemingly simple issue like the value of corn can have large effects on the best use of a given parcel of land. Scenario analysis can reveal the costs, benefits, and sensitivities of specific policies.

Scenario analysis can demonstrate the landscape effects of policy decisions, and how applying the highest and best use principle can help clarify the best decision to make—it is a decision support tool. Dr. Wainger said benefit indicators and monetary values are useful for determining landscape-scale effects. Overall, she concluded, ideal scenarios are visions to be achieved, and by incorporating modeling tools, economic incentives, laws and policies, and social pressures, these ideals can be achieved.

Murray Hitzman, Charles F. Fogarty professor of economic geology at the Colorado School of Mines, discussed the relationship between multi-resource analysis (MRA) and mineral resources. Mineral resources are fixed in space unlike other natural resources, such as wildlife or water resources that move from one location to another. A challenge with developing mineral resources is the large volume of rock or soil required to be removed to extract a small amount of the mineral, he noted. For example, copper mines produce copper from rock with between 0.5 to 1.5 percent copper (5,000 to 15,000 parts per million Cu per metric ton of rock).

Dr. Hitzman said that the extraction of mineral resources is not created equal with different minerals having very different physical and temporal footprints. The Olympic Dam copper mine in Australia is one of the largest copper mines in the world; however, it is underground and has a land surface disturbance of only 13 square kilometers. In contrast, the Gold Strike gold mine in Nevada is an open pit mine with a land surface disturbance of 130 square kilometers. A large open-pit copper mine in Utah was in operation for over 100 years, whereas a copper and gold mine in Wisconsin was in operation for only eight years. Mineral resources vary in their value as a commodity. Some minerals can be very high in value, such as diamonds or gold, and are transported throughout global markets hundreds or thousands of miles from where they are mined; however, other resources are lower in value, such as limestone and sand, and are transported only tens of miles from where they are mined.

Minerals can be economically important commodities at one point in time, but may lose that importance over time. Platinum was not considered an economically important commodity prior to the 1970s when catalytic convertors for automobiles were created. Platinum very quickly became a critical material for the United States. The global spatial distribution of minerals varies as well. For example, platinum deposits are not available in the United States, and are mostly located and mined in South Africa. Copper and zinc deposits, however, are mined in many countries, including the United States. Thus, different minerals have different supply risks. The National Academies of Sciences, Engineering, and Medicine issued a report[3] on the criticality of minerals, which defined criticality based on its importance and supply risk (Figure 4-5).

Dr. Hitzman said that a prospective analysis would determine if a landscape has the potential for a mineral resource, and a reactive analysis would determine the impact on the landscape of developing that resource. Reactive analysis, he said, determines whether an impact on a landscape is compatible with other services provided by the landscape (e.g., water supply, recreation, energy). The analysis is based mostly on environmental impact statements and environmental impact assessments of a site prepared by the private sector; however, regulatory land and environmental agencies must consider all potential impacts over a long time scale. A prospective analysis, said Dr. Hitzman, is more similar to an MRA because it requires a wide scope and is in line with the type of studies government agencies conduct.

This analysis would address finding geologically favorable areas for mineral resources, the likelihood of economically significant amounts, and the potential

---

[3]National Research Council (NRC). 2008. *Minerals, Critical Minerals, and the U.S. Economy*. Washington, D.C.: The National Academies Press.

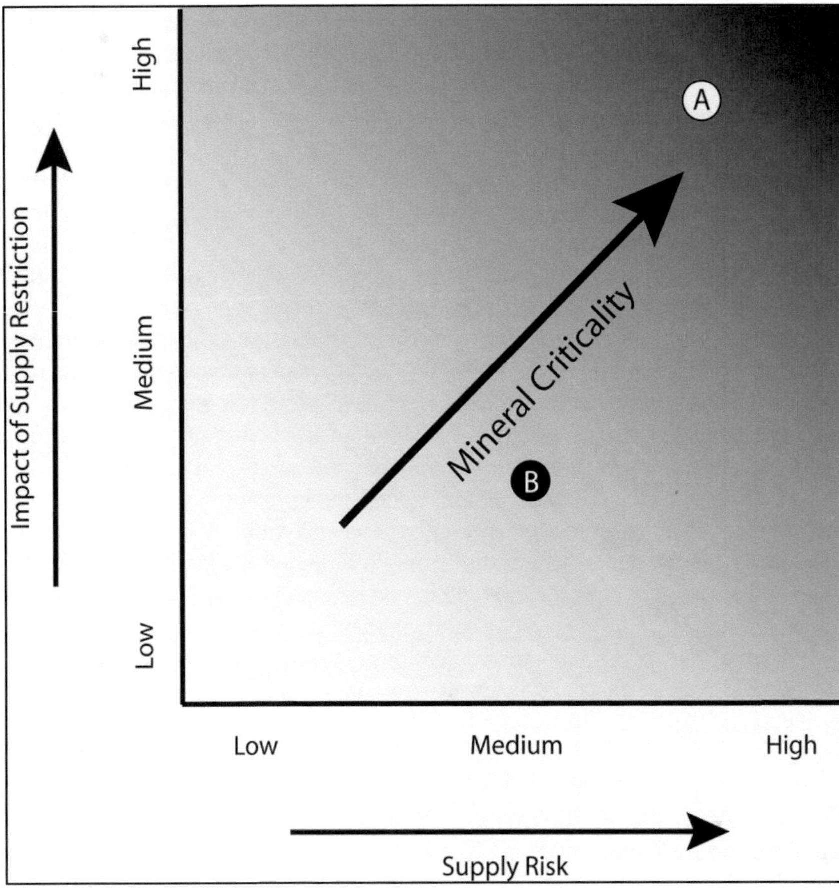

FIGURE 4-5 The criticality of a mineral as a function of its impact on supply restriction and supply risk.
SOURCE: NRC (2008) and Murray Hitzman, Presentation, National Academies of Sciences, Engineering, and Medicine Workshop, June 2, 2015, Washington, D.C.

environmental and/or social impacts of exploring and extracting the resources. The analysis would also address the economic risk and benefits of a private sector entity investing in the extraction of those resources in a way that minimizes impacts on the landscape. A private sector entity could use such information to determine the potential return on an investment and whether it is great enough to warrant the initial investment and potential risk.

Dr. Hitzman said that the role government has in conducting MRA for mineral resources would be to delineate areas with mineral deposits—such as is currently done through USGS geological mapping—classify land where mineral resource development can occur, and identify areas where other land-use considerations preclude mineral development. Dr. Hitzman provided an example of government information aiding mineral development from Gawler Craton, South Australia. The government derived geophysical (magnetic and gravity) surveys, which led to the development of numerous geophysical targets. Such maps are two-dimensional and provide an image of minerals in the subsurface. These government-derived geophysical maps illustrate where the greatest mineral potential is, and are critical for prospective analysis by the private sector. Dr. Hitzman said that development of different products by government agencies is helpful to the mineral industry, including ore deposit models, geologic maps, and geophysical data to help determine geographically favorable areas. Also, he added, data are critical for developing these products, and landscape and commodity criticality data are needed to guide decision making in developing resources.

Dr. Hitzman commented that MRA should ask whether or not a landscape has the potential for societally beneficial mineral resources. Societally beneficial aspects can include the criticality of the deposit and infrastructure improvements, such as roads. MRA would also assess the economic value or benefit needed to overcome possible adverse impacts to the environment or landscape. Mineral resources are fixed in their location, which often results in conflict among stakeholders with a decision to extract a mineral resource. This conflict, Dr. Hitzman said, is why MRA is needed—to provide more informed decision making.

Dean Urban, professor of landscape ecology at Duke University, presented on the challenges that biophysical heterogeneity and social factors pose when managing landscapes and on multi-criteria decision frameworks for informing decisions on landscape management. He presented the software program Portfolio, which was developed for a Nature Conservancy (TNC) project assembling eco-regional portfolios.[4] The software uses what he termed a greedy algorithm that allows users to assess conservation practices for different sites. A greedy algorithm follows a problem-solving heuristic of making the locally optimal choice at different stages with the hope of finding a global optimum. The TNC project assessed habitat patches in terms of biodiversity, species richness, species rarity, habitat geometry, area effects, edge effects, and threats from development. Dr. Urban said that the program was originally written nearly 15 years ago to teach graduate students about multi-criteria decision analysis, and he is encouraged to see the traction multi-criteria decision analysis has gained especially with leadership from USGS.

One lesson learned over the years, he said, is that practitioners are not interested in an optimal solution, but rather are interested in reaching better solutions. A portfolio of sites in the program is developed by assessing many different sites with many different capabilities. TNC, Dr. Urban said, deliberately used the term

[4] Urban, D. 2002, in Learning Landscape Ecology: A Practical Guide to Concepts and Techniques, G. E. Sarah and M. E. Turner, eds. Berlin, Germany: Springer.

"portfolio" because, like a retirement portfolio with different financial investments, there are different risks and potential returns for each site. Individual sites are not being optimized, but rather a whole set of sites are optimized. The Portfolio software package is a decision framework with two pieces—an objectives hierarchy and a path model. The objectives hierarchy is a statement about what is being measured in the system and the targets that are set. In the TNC example, the target was to preserve biodiversity. There were different objectives—diversity, patch geometry, and landscape context. Each objective is indexed by empirical indicators, such as richness in biodiversity. The objective hierarchy is coupled to the path model, which is a statement about how actions will affect those objectives. It is a causal chain that links actions to outcomes, and is also intended to capture indirect effects that lead to impacts on non-target resources.

For many ecological applications, the links are implicit in the path model. For example, "area" as an indicator invokes the species-area relationship, geometry (e.g., edge/area indices), edge effects (e.g., forest birds in eastern United States), and connectivity. Dr. Urban presented an example of landscape-scale management of forests to reduce the risk of catastrophic fires. A simple path model would posit that managing a forest understory would reduce catastrophic fire risk. Thus, changing the forest structure by managing lighter fuels and fuel bed connectivity should result in a reduction in fire risk. Extending the path model allows a landscape manager to consider other factors that would be affected by managing the forest structure in this way; however, there is much work needed in better understanding all the connections. There has been extensive work on fire management in the western United States, for example, but little progress made on using wildlife habitat suitability models to examine what that management means for biodiversity.

Dr. Urban said that extending such applications to ecosystem services requires extending the ecologist's view of ecosystem structure and dynamics to the supply of ecosystem services to stakeholders. To do this, stakeholders need to be identified and supplied those services, which is captured using a human well-being index. The index is quantified by monetization. There is a need, he said, for more datasets and models that would help identify stakeholders and their preferences for how resources should be managed. There is also a need for more data on ecosystem services at large spatial scales. A crucial step in extending application to ecosystem services is the service area analysis, which identifies benefits to stakeholders. Dr. Urban provided an example from the Triangle Land Conservancy in North Carolina, which identified stakeholders that most benefited from access to nature reserves held by the land trust but available to the public. The service areas were analyzed in increments of 5-, 10-, 15-, and 30-minute travel times from seven different reserves. The analysis found that the reserves were more accessible to white and wealthier stakeholders living in proximity to those reserves. Dr. Urban said that the geographic information system (GIS) used to conduct such an analysis was originally developed, to some extent, to do such an analysis but needs to be developed further so that it helps make land management and stakeholder decisions. For example, he said, GIS needs to be better integrated into planning to determine where to best site public services, such as fire stations or hospitals, which optimizes access to stakeholders.

One challenge, Dr. Urban noted, is there is no single overarching model that can pull together all the different components required for a full assessment of management options and link them to public well-being. The Federal Guidebook on Managing Ecosystem Services, launched in December 2014, provides alternative matrices of management scenarios and ecosystem systems.[5] The benefits garnered from the intersection of the services and management scenarios fill in the matrix. Constructing such a matrix allows the manager to quickly assess the value of different benefits derived from alternative scenarios. He provided an example of forest thinning from the guidebook, which positions various models (i.e., fire behavior, species distribution and population, and air plume models) between key elements of the path model (i.e., forest structure, fire behavior, species habitat, species population, and air quality). He emphasized that there are multiple models invoked throughout the assessment to estimate the production of services and value of those services. There is no single model, however, that can pull together all of the individual components to elucidate the ecosystem service values of fire management and the effect the services ultimately have on social well-being, such as the effect on healthcare costs from changes in air quality. Dr. Urban concluded that the chain-of-custody approach with multiple individuals conducting one part of the analysis and then passing it along to others leads to errors throughout the assessment process; there needs to be improved stewardship of the process.

In the question-and-answer session, the panelists responded to a question about overcoming the challenge of providing ecological outputs from models that are then used as inputs to economic models. Dr. Urban said that as an ecologist working with economists he finds it rewarding to produce model outputs in a format that economists can readily use. He added that although it may seem a simple process, it is challenging because it requires ecologists to shift their thinking about how they develop and produce model output. Dr. Wainger said it is a continuing area of work and that often the models do match directly, but there are ways to work around incompatibilities. For example, using benefit indicators suggests that not all data from models will be able to fit together for an evaluation, however, taking into consideration the public's concerns and preferences is still possible to estimate relative differences in benefits. Dr. Hitzman added that the data from any one person's work should be able to be used in other models and understood by other professionals. It would, however, require many professionals to change the type of output they produce.

Another question posed to the panelists was about how to connect dis-

[5] National Ecosystem Services Partnership. 2014. *Federal Resource Management and Ecosystem Services Guidebook*. Durham: National Ecosystem Services Partnership, Duke University. [Available at: https://nespguidebook.com].

parate types of natural resources. The panelists had spoken about commodity criticality, optimization, and MRA as themes in their talks. The question posed was if commodity criticality could be integrated into MRA, and similarly, could commodity criticality also encompass optimization criteria. Dr. Wainger replied that performance, costs, and benefits can all be incorporated into an optimization model. The scale of the analysis has to be in line with the questions being asked. For example, she said, in her mountaintop mining case study there was a solution provided by the optimization model for a given question within a set area; however, if that area is broadened out to include other regions, a different solution may be optimal. Dr. Urban emphasized that methods used in a multi-criteria analysis are well suited to incorporate risk and uncertainty. Dr. Hitzman agreed that tools already exist to bring optimization and risk assessment methods into an assessment of commodity criticality.

A final question related to how to address error propagation in modeling; even with ideal data, a model will have a range of errors associated with the output. As the output from one model is used in other models, the output of those models, and their associated errors, are then used in more models. There is an accumulation or propagation of errors throughout the series of model outputs. Dr. Urban responded that especially when working in fairly complicated models, error propagation can occur very quickly. It is critical, he said, to be transparent about how much uncertainty there is in a model so that it is clear to stakeholders. He added that by being transparent and knowing where in the model there is error propagation allows an agency conducting the modeling to invest in resources necessary to minimize that uncertainty. This is a key component of adaptive management. Dr. Wainger said that error propagation can often be hidden, but can be minimized by ensuring the right level of detail is included in models used in a decision making context. It is useful, she added, to use a more conceptual model when there is more uncertainty in the data, and that it is not critical to have a robust decision; there can be a range of methods used to arrive at a decision without using complex models and conducting a rigorous uncertainty analysis.

5

Methods for Evaluating Scenarios: Reconciling Quantities and Values

Richard Bernknopf, research professor at the University of New Mexico, presented a collaborative project among his research group, the U.S. Geological Survey (USGS) and Sandia National Laboratory. The Net Resource Assessment (NetRA) is an assessment of the societal tradeoffs and impacts on ecosystem services with a resource extraction or economic development activity. The assessment is driven by the USGS Energy and Minerals Science Strategy and Department of the Interior Secretarial Order 3330. The goals of the USGS Energy and Minerals Science Strategy aim to provide inventories and assessments of energy and mineral resources, and to understand the effects of energy and mineral development on natural resources and society.[1] Additionally, Secretarial Order 3330 states that "the Order will ensure consistency and efficiency in the review and permitting of new energy and other infrastructure development projects . . . while also providing for the conservation, adaptation and restoration of our nation's valuable and natural and cultural resources."

Dr. Bernknopf said that the NetRA is an analytical component of a multi-resource analysis (MRA) and is accompanied by a decision support tool. It takes account of current societal decision-making demands and is an expansion of natural resource assessments that include ecosystem services. The current USGS resource assessments, he said, can be broadened by developing a Decision Support Tool (DST)-oriented conceptual framework that assesses the benefits, costs,

[1] Ferrero, R.C., Kolak, J.J., Bills, D.J., Bowen, Z.H., Cordier, D.J., Gallegos, T.J., Hein, J.R., Kelley, K.D., Nelson, P.H., Nuccio, V.F., Schmidt, J.M., and Seal, R.R. 2013, U.S. Geological Survey energy and minerals science strategy—A resource lifecycle approach. U.S. Geological Survey Circular 1383–D, 37 p. [Available at: http://pubs.usgs.gov/circ/1383d/circ1383-D.pdf].

and societal tradeoffs associated with the collocation of natural resources and ecosystem services. The DST-oriented conceptual framework also provides a tool for policy analysis with simulation capabilities. The tool for policy analysis is a system dynamics model that allows for the simulation of multiple resource development scenarios and provides a comparative analysis of the outcomes of different policies and practice.

The NetRA is initialized with USGS natural resource and ecosystem data, and integrates multiple collocated resources to generate usable development scenarios, he explained. It contains both spatial and temporal components to estimate the net societal benefits for a scenario of regional natural resource development. It also applies the DST systems dynamics model to evaluate specific resource development scenarios resulting from landscape conversion. Dr. Bernknopf said that there was a site selection process with USGS to establish a set of formal criteria to evaluate active or the potential development of a resource. The criteria included collocation of ecosystem services and tradeoffs of interest, any public land involved, available data, and potential decisions for the Bureau of Land Management (BLM). Of five candidate sites (the Greater Green River Basin, San Juan Basin, Piceance Basin, Bakken Shale, and the Uranium Time-Out Area), the Piceance Basin in northwestern Colorado was selected, in part, because it was less controversial than the other sites.

Resources in the Piceance Basin are intertwined in a way that extracting any one resource would have an effect on the overall ecosystem services. Natural gas extraction by hydraulic fracturing was chosen as the resource assessment unit. There are diverse opinions among stakeholders on hydraulic fracturing and the potential for development in the Piceance Basin; however, there is currently no natural gas being extracted. The group decided that the ecological assessment units defined would have geographic boundary, surface and subsurface components, and clearly identifiable and measurable ecosystem services. Habitat fragmentation, water quality, and elk migration were chosen. In addition to the complexity of the resources, Interstate 70 cross-cuts the study site and the Colorado River is down slope from the site. Comparing elk migration patterns with shale borehole locations and leases reveals that the elk migration corridors are collocated with the gas resources. Any development of the gas resources would have an impact on the elk regardless if the migration patterns were permanent or only seasonal.

Dr. Bernknopf said that the DST would incorporate multiple disciplines including biology, ecology, geology, hydrology, economics, and engineering; integrate natural resource and economics value concepts, methods, and data; and couple models and scientific data with market and nonmarket prices. Additionally, it would estimate the net societal benefits of all natural resources developed and preserved. It can be applied at multiple scales (e.g., interregionally or intraregionally), used temporally as well as spatially, and operated in multiple regulatory constraint frameworks. The assessment aims to find societal benefits impacted by environmental impacts, development costs, and natural gas production. Scenarios can be developed to demonstrate how changing one component affects the others. For example, by focusing more on development, the environmental impacts and net societal benefits can be assessed. This is done through a series of models that form a network of stocks and flows—the stocks are the resources in place and the flows are the development through time and space. He said that assessing feedbacks between the stocks and flows is also important.

Dr. Bernknopf presented the production function model for the extraction of natural gas by hydraulic fracturing. There are multiple components in the model, however, that are still being developed and that do not have data available. Additionally, there are sub-models to the production function model. The geology flow rate sub-model captures characteristics such as number of wells, depth of wells, and permeability of the rock. It captures the characteristics that are used in geological assessments to develop a flow rate, which is a well-known indicator used in practice. This geologic sub-model feeds into a hydraulic fracturing sub-model, which assesses the development of the resource with number of wells, distance of horizontal expansion, pad density in the area, and number of roads. This assessment then feeds into a cost sub-model to estimate the total cost of extraction. He also presented an ecological model that would assess the percentage of elk population that would be displaced and how the migration patterns would shift. Inputs into this model included development characteristics, such as road sizes, road density, pad sizes, and pad density in the area. The outcomes from the different models are condensed in the RAU, which quantifies the fuel or mineral development, and the EAU, which quantifies the environmental and ecosystem impacts (Figure 5-1). Reserves and resources can be derived from the RAU and water quality and habitat fragmentation can be derived from the EAU—both are ultimately used to assess the total social benefit of development.

Dr. Bernknopf said that the NetRA project is in the proof-of-concept stage to demonstrate the conceptual framework's feasibility. To do this, however, there are still data for inputs and issues with model compatibility that need to be solved. Several data gaps exist, including natural resource stocks, engineering economics for resource extraction, biophysical and ecological data for ecosystem services stocks, market prices, regulations, and nonmarket values. The output of the proof-of-concept, he said, will be a modeling framework that demonstrates the functionality of the decision support tool. The decision support tool will be a probabilistic model that will provide a distribution of outcomes, expected values, and uncertainty. The conceptual framework and DST will attempt to meet development needs (e.g., natural gas resources) and ecosystems needs (e.g., limit impacts on terrestrial ecosystems) by objectively providing a probability distribution of outcomes that allows the decision maker to compare one scenario to another.

Karen Jenni, founder and president of Insight Decisions, is developing an integrated MRA with the USGS for a defined geographic region in the Powder

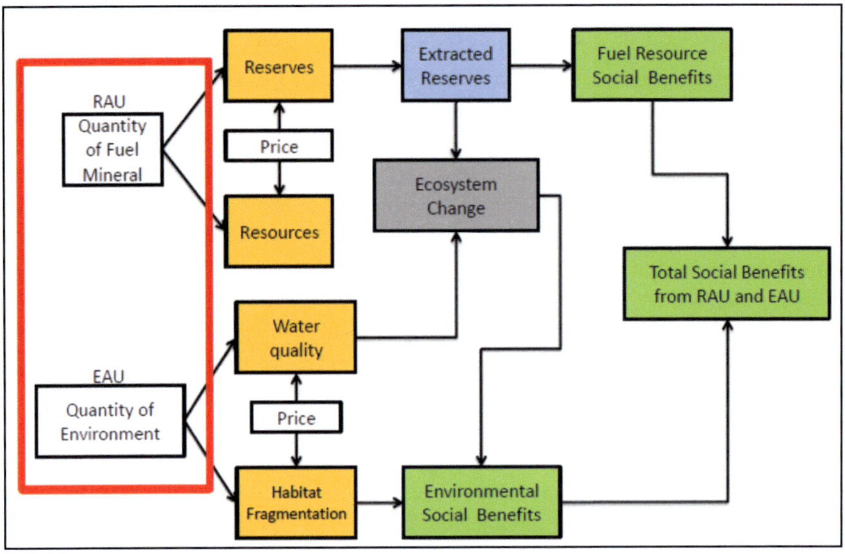

FIGURE 5-1 Outcomes from different models are condensed in the RAU and EAU, which are used to assess the total social benefit of development.
SOURCE: Richard Bernknopf, Presentation, National Academies of Sciences, Engineering, and Medicine Workshop, June 2, 2015, Washington, D.C.

River Basin in Wyoming and Montana, which she highlighted in her workshop presentation. She is evaluating how to combine an analysis of existing resources under current conditions as well as under different future scenarios. An MRA, she said, consists of three main components: integrated information on the current status of multiple natural resources (including ecosystem services), models describing the interrelationships among collocated natural resources, and analyses evaluating the impacts and tradeoffs to the natural resources in biophysical and socioeconomic terms (Figure 5-2).

For the Powder River Basin, Dr. Jenni conducted a regional inventory and assessment of multiple natural resources, and included summary-level descriptions of resources suitable for different audiences. She also has conceptual models and analyses that address the relationships between the natural resources in the region; however, she is still early in the process of developing the models. One area of importance, she emphasized, is the resource valuation and economic implications of changes in resources under different future scenarios. An important component for stakeholders is a "portal" to access information (or multiple "portals" but with resource information available through each portal). The current interface is a Geographic Information System (GIS)-based system that provides detailed resource information until ultimately a source document,

Component of MRAs	PRB MRA elements
Integrated information on current status of multiple natural resources (including ecosystem services)	A regional inventory and assessment of multiple natural resources, including summary-level descriptions of the resources suitable for different audiences
Models describing the interrelationships among collocated natural resources	Conceptual models and analyses that address the relationships between the natural resources in the region
Analyses evaluating the impacts and tradeoffs to the natural resources in biophysical and socioeconomic terms	**Resource valuation and economic implications of changes in resources under different future scenarios**
	A "portal" to access the information (or multiple "portals" but with all resource information available through each portal)

FIGURE 5-2 The three main components of a multi-resource analysis (MRA).
SOURCE: Karen Jenni, Presentation, National Academies of Sciences, Engineering, and Medicine Workshop, June 2, 2015, Washington, D.C.

such as a USGS report, is provided. This was a critical feature to USGS and to other stakeholders—maintaining transparency and traceability of data used in the models.

Dr. Jenni described effective approaches for measuring and comparing values provided by multiple resources within a landscape-based approach. Involve decision makers and stakeholders early and often, she said, and start with the objectives of the decision makers or stakeholders, and allow them to specify their values. She also suggested acknowledging difficult-to-quantify values, and to the degree possible, separating "technical" questions (e.g., biophysical outcomes) from "value" questions. She described an example of the efforts with the Landscape Conservation Cooperatives in the Pacific Northwest that engaged tribes and First Nations in Canada to recognize traditional ecological knowledge and values that were important to their tribal communities. Although some of the traditional knowledge was difficult to value, it was still important to recognize. The values of first foods, for example, were challenging. Salmon is an evident food that had value assigned to it, but there were other first foods important to the tribal stakeholders that did not fit into an evaluation framework. Lastly, she said, it is important to maintain transparency throughout the process when aggregating values, with no "black boxes." In engaging stakeholders in the Powder River Basin, she said many are interested in knowing more about the whole process and even in how modeling is conducted. It is important but challenging to engage stakeholders at such a level of granularity within a project conducted on a large scale assessing multiple resources.

Dr. Jenni stated that decision analysis as a quantitative methodology brings together three important characteristics. The first is that decision analysis provides a process and structure to addressing a resource management issue. For example, decision analysis can help define the scope that would be addressed under a multi-resource assessment. Clearly defining the decisions to be supported and outcomes that are important to decision makers will limit the overall scope, as well as scope-creep, in a resource management project. Bringing many researchers together on a project can result in more information than what is needed to address a particular issue; it is important to keep focus on a set of outcomes that decision makers have expressed are important.

Second, decision analysis offers many tools for quantifying uncertainty. It is used extensively in modeling efforts to reduce overall complexity of models. Expert assessment can be used when not every step in a process is modeled, meaning that judgment can be used to capture what is known and not known for modeling. Uncertainty is used to summarize what is known in the model and to move beyond the unknown steps. It helps to explain how certain modelers are about their outcomes and to infer what may happen if they are wrong.

Third, decision analysis allows for multi-attribute or multi-criteria approaches. Dr. Jenni said that often stakeholders do not want separate outcomes for different priorities combined into just one value, but instead want to see how different decisions may impact each of the priorities separately. It is important for stakeholders to have common goals so that decisions can be identified to best reach those goals. Such agreement is a challenge, she said, especially when decisions are on a large, landscape-based scale.

Dr. Jenni said that the steps necessary to link biophysical quantities to values, and to ensure biophysical measures or models are consistent with quantification, start with decision makers' and stakeholders' desired outcomes. It is important to start with understanding what decision makers and stakeholders want and then working backwards in the process. Identify the values linked to a decision framework and to scenarios of interest, and then identify biophysical measures that relate to those values. Also, she said, it is important to document the whole decision analysis process, which will also limit the over-modeling that often occurs.

Robert Johnston, director and research professor of the George Perkins Marsh Institute at Clark University, discussed methodologies for linking quantities and values, which he described as very challenging to do empirically even if simple to do conceptually with existing frameworks. Dr. Johnston started with a few key points. A landscape-based analysis is only as good as its weakest link, he stated. For example, if a decision support tool is used, but the science incorporated into it is not very good, then the entire analysis can be considered limited. Also, there are often arbitrary constraints and expectations on an analysis. Often, an analysis is approached with an idea of how an issue would be incorporated into a decision support tool, a requirement to assess values across an entire landscape, or with a specific way in which an agency would address an issue due to cultural or political reasons. Such constraints often make the analysis more difficult and do not necessarily comport with how systems would actually behave.

Dr. Johnston said economic evaluation as a multi-resource decision framework is designed to evaluate tradeoffs across multiple different outcomes and is explicitly designed to link the different biophysical outcomes to values. One advantage of economic valuation is that it provides a formal quantitative approach and enables internally consistent and generalizable results. Expressing outcomes in dollar values allows them to be more easily compared in a meaningful way and provides a consistent metric to measure changes in social welfare; however, the evaluation of tradeoffs does not have to be expressed as dollar values but can be other units of measure. Another advantage of expressing outcomes in dollar values is that it measures values realized by the general public, which may not be aligned with the values held by policymakers or other stakeholders. Market values and ecosystem service values, Dr. Johnston said, are subsets of economic values or benefit cost analysis. Determining tradeoff ecosystem-service values versus resource-extraction values versus market values are all the same to an economist, he said. They can be viewed as goods and services, but there are differences in how they are measured. Economic values are only one part of the assessment, and do not dictate the direction decision makers will take. They are used with other information, including other values measure using other approaches, in the overall decision-making process.

Dr. Johnston provided as an example a case study of Kitts Hummock, Delaware, which focused on eroding beaches (Figure 5-3).[2] The management questions pertained to spending money to maintain beach or allowing various types of retreats of the beach. Considerations included habitat for shorebirds and horseshoe crabs, as well as housing that lined the beaches. The first step in addressing the management issue was to consolidate the considerations in the valuation process. This included compiling the cost of sand, fill, and demolition of structures, buying out homes from homeowners under retreat scenarios, recreational benefits, property transfers, and housing services. Values can be assigned to all the considerations and a cost and benefit analysis emerged. When doing an objective cost and benefit analysis, he said, the outcome can sometimes be surprising. For example, the option that provided the most benefit was to allow the beach to erode naturally.

---

[2] Johnston, Robert J., Mahesh Ramachadran, and George R. Parsons. 2014. Benefit Transfer Combining Revealed and Stated Preference Data: Nourishment and Retreat Options for Delaware Bay Beaches. [Available online: http://works.bepress.com/cgi/viewcontent.cgi?article=1041&context=george_parsons].

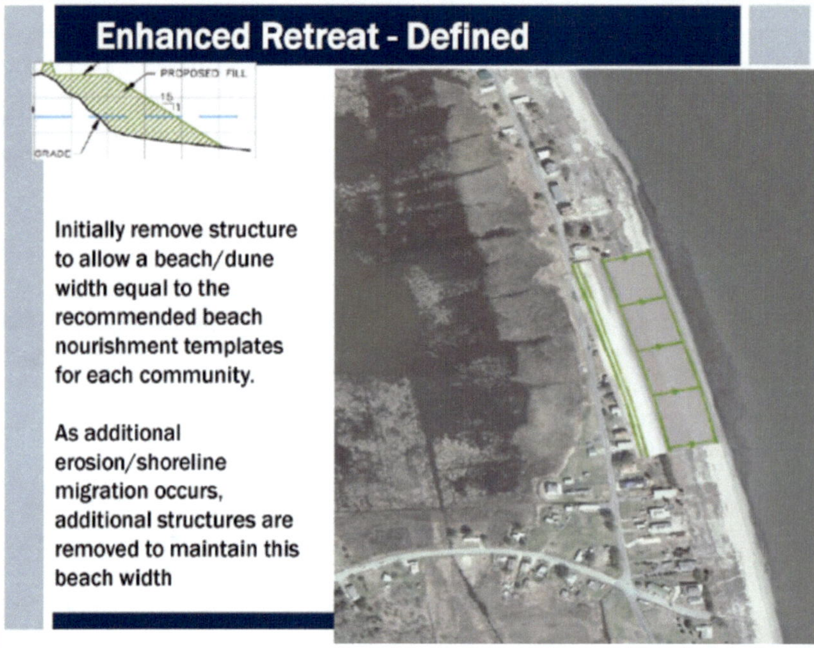

Scenario	(A) Sand, Fill and Demolition (PV, $mill)	(B) Housing Acquisition Payments (paid by State) (PV, $mill)	(C) Housing Acquisition Payments (received by property owners) (PV, $mill)	(D) Recreation (PV, $mill)	(E) Housing Services (PV, $mill)	(F) Reduction in Add'l Flood and Erosion Damages (PV, $mill)	(G) Net Benefits (PV, $mill; sum of A through F)
Nourish-ment	-$61.1	-$0	$0	$16.1	$18.2	$2.7	-$24.1
Basic Retreat	-$0.5	-$61.3	$61.3	$10.8	-$43.1	$3.0	-$29.8
Enhanced Retreat	-$4.5	-$149.1	$149.1	$10.8	-$130.9	$10.6	-$114.0

FIGURE 5-3 An assessment of management options to address eroding beaches in Kitts Hummock, Delaware.
SOURCE: Robert Johnston, Presentation, National Academies of Sciences, Engineering, and Medicine Workshop, June 2, 2015, Washington, D.C.

It is standard practice, he said, to compare economic values provided by diverse market and nonmarket resources across landscapes. Substitutions of resources (tradeoffs) must be realistic and plausible. This becomes very challenging when cultural values of First Nations or native communities are involved. There are many perceived challenges when doing economic analysis, because often people have an objection to the idea of monetization and the degree to which it is applicable—a fear of economic imperialism. There is also a misunderstanding and misapplication of methods, particularly with mapping and decision-support tools. This can be true with benefit transfer methods where a value is calculated in one location and then applied in another.

External constraints are also a challenge to the economic valuation process. Often researchers impose their own constraints on the analysis. For example, he said, a researcher may set out to map value and needs to determine a value for every resource. The challenge is that the value of a particular resource in a given area is a factor of everything else around it and meaning is lost when a static value is attached to it. There are also subjective distinctions made between values. He said that different agencies may have enabling legislation or regulations that state only certain types of values can be considered. This can lead to distinctions between valuation tools or a push for one-size-fits-all decision support tools; however, he said, such arbitrary distinctions are often not supported by the scientific literature.

Typically, Dr. Johnston said, only a subset of primary values can be measured empirically. Economists often categorize values as either market or nonmarket values (Figure 5-4). Nonmarket values are often associated with observable behavior and nonuse values, such as the values of existence and for passing on a resource to future generations. Dr. Johnston noted that although the diagram for market and nonmarket values is fairly simple, it is well established. The diagram does not, however, contain ecosystem service values explicitly.

Ecosystem service values are not often a useful distinction from other types of values. Rather, he said, it is more useful to determine what biophysical elements affect human well-being, and then to label the element as an ecosystem service value if it helps communication. Often though, such arbitrary distinctions can confuse rather than help the analysis. One of the challenges is how to link biophysical models with economic ones, which can be done when natural scientists measure the right elements for valuation.

Economists do not measure the values of total ecosystems, but rather the value of changes in specific measures. It is important to determine what to count as a basis for valuation (i.e., developing causal chains or means-ends diagrams). This is done by first identifying the beneficiaries by fiat (e.g., political jurisdiction) or by analyzing where values exist (e.g., economic jurisdiction). Then, management or policy actions being considered are linked to biophysical changes in the landscape while also accounting for human behavioral changes as necessary. The benefit-relevant indicators along the chain are then identified (e.g., the

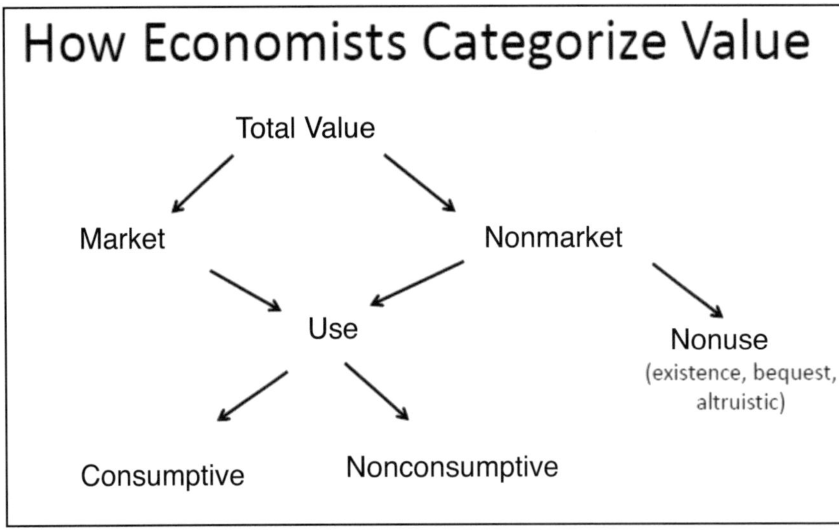

FIGURE 5-4 How economists categorize values as either market or nonmarket.
SOURCE: Robert Johnston, Presentation, National Academies of Sciences, Engineering, and Medicine Workshop, June 2, 2015, Washington, D.C.

fundamental things that people value, such as recreation, hunting, or property value). Once the indicators are identified, they should be qualitatively described to represent what people value without ascribing an actual value, after which economic valuation or another multi-criteria decision analysis tool can quantify the social value. This links biophysical changes to primary changes in social welfare.

The panel was asked a question about the user interface of decision support tools and modeling, and if there is a way to make them more interactive with non-modelers. The questioner said that often when there is a conflict on the landscape, the stakeholders want to understand a decision support tool that is being developed to address the conflict, but due to the complexity of the modeling, they come to see it as a black box. Stakeholders want to be part of the decision-making process, but there is concern their values are not being incorporated into the process. Dr. Bernknopf explained that they are trying to develop a decision support tool that allows all users to create a range of scenarios and realize outcomes without having to understand all the equations and statistics incorporated into the model.

Dr. Johnston added that uncertainty with models and the stakeholders' comprehension of that uncertainty provides an ongoing challenge. The need to provide simplified information needs to be balanced with the need to provide accurate information. For example, when a decision support tool results in a map, there is a tendency for stakeholders to take the map as truth and not understand the extent of uncertainty underneath the representation of data. Dr. Jenni said that there is value in using simplified, cartoon models to communicate information. Often when there are multiple disciplines involved in an assessment of a large area, such as the Powder River Basin, the modeling becomes too complicated when all interests are incorporated and cannot be explained in a more simplified way. Dr. Bernknopf added that a decision support tool can assess a range of outcomes, which helps to address some of the uncertainty.

The panelists were asked about the complexity of models and if there were any attempts to reduce the amount of complexity to determine the least amount of information needed to successfully use the models presented. It was noted that the more complexity the models contain, the fewer stakeholders would be able to use them, and the less successful implementation would be. Dr. Bernknopf replied that they are working with a range of different types of scientists and trying to accommodate everyone by keeping the models manageable for all involved yet scientifically robust. Carl Shapiro, director of the Science and Decisions Center at USGS, commented that as the sponsor for Dr. Bernknopf and Dr. Jenni's projects, they aimed to do a proof of concept of two multi-resource analysis projects in two places to see if they could be developed in a way that was relevant and useful to decision makers. Dr. Shapiro said that they are not yet at the stage of trying to eliminate complexity from the model as they are still early in the process of modeling to see if the efforts are feasible.

A related question to the panelists was if the decision tools being developed are able to factor in variances in the decision space associated with different resources. The example given was for resources that had explicit trust laws that limited the number of possible decisions. Dr. Bernknopf responded that they are trying to find constraints over a large region, and to determine the probability of the amount of resources capable of being extracted in that region. They have not yet been able to address constraints at such resolution as specific regulations that would protect a wilderness area. Another participant commented there are two conflicting goals—the development of a sophisticated decision support tool that aggregates a tremendous amount of information for a decision maker but that is also simple enough for a nontechnical user to manipulate and understand. The participant said that there is a significant challenge in being able to have the sophistication necessary in a decision support tool while also keeping it simple enough for all stakeholders to use.

The panelists were asked about how to address stakeholders when they assign different weight to different values in a multi-resource analysis context. The participant explained that when stakeholders are provided with a range of scenarios with different outcomes and are asked to make policy or management decisions, the challenge is that they value different attributes of the scenarios and could potentially become gridlocked in making a decision. It may be useful, the participant added, to provide the multi-stakeholder decision with all the outputs disaggregated by values to compare against what stakeholders think would happen in the absence of a management decision.

Dr. Bernknopf responded that they hope to develop a decision support tool that would essentially allow for that process by providing stakeholders with virtual dials that they can use to change inputs to see how the outcomes change accordingly. By allowing all the stakeholders to create a range of outcomes over multiple scenarios, a distribution of values is provided. He said it is a probabilistic approach to the decision-making process. Dr. Jenni added that it is important to allow different stakeholders to provide their input and express their tradeoffs in the modeling process. She said there is also a challenge in identifying who the decision makers are in a large, landscape-based analysis. Many stakeholders are interested in the outcomes of scenarios, but they are not necessarily the decision makers that need to use the decision support tool.

Dr. Wainger said that developing models and scenarios is aimed to influence specific decision makers, such as landowners. The goal may be to have them install a structure, such as a buffer strip to protect water quality. The question she posed was if landowners needed to understand all the benefits they are providing to society or if they only need to understand their own personal benefits or own opportunity costs. Much of that discussion, she added, is about how much detail or data is needed for the analysis when many of the attributes of the analysis are fundamentally incompatible with each other. The other questions she posed were about when quantities are useful and their limitations for helping understand tradeoffs. Researchers often quantitatively analyze the attributes of a decision analysis but are still unable to resolve tradeoffs, which results in negotiated solutions that may or may not be based on the data available. Dr. Jenni provided an anecdote from a recent meeting where a scientist presented on a decision analysis preformed with stakeholders that resulted in a decision not based on the analysis but on the stakeholders' own perspectives. Decisions, she said, are often not based on facts or data but on the values held by the decision maker.

Dr. Hitzman added that there are factual-based (data-based) models and value-based models. He said that there is a challenge when it comes to the decision maker using a factual-based model, because ultimately decisions are often political. He said that it is critical to consider the political reality when building decision support tools, although it may not be used until the model or decision support tool is communicated to stakeholders. The communication step, he added, needs to be separated out from the model itself. Dr. Shapiro provided perspective on the complexity of the models discussed. He said that one of the reasons for developing complex models is due to a relatively complex set of issues when conducting a multi-resource analysis with multiple stakeholders over a large landscape. The process begins by moving from a single resource analysis to trying to understand what happens to other resources in the same region, and then ultimately moving to a valuation step. There will be a necessary tradeoff, he said, between how much information will be available to inform a model and the complexity of a model so that it is not so complex that a decision maker is unable use it.

6

Multi-Disciplinary and Cross-Agency Synthesis

Mark Schaefer, a global fellow at the Woodrow Wilson International Center, presented on how to implement a multi-resource analysis (MRA) within a policy context. He commented that there are many terms used by researchers that often describe the same elements of an analysis. There is value in clearly communicating what is being advocated to a nontechnical policymaker. Clearly communicating the basic characteristics and benefits of MRA begins with agreeing on a set of easily understood, fundamental facts describing MRA and why it will advance resource conservation and sustainability (i.e., devise and reiterate the "one-minute elevator speech"). Multi-resource analysis provides a mechanism to define goals for resource use and conservation at a regional scale, set clear priorities to attain those goals, further sustainable resource use and planning across landscapes and regions, and provide a cost-effective approach.

Dr. Schaefer said that there is value in pursuing intersectoral collaborative opportunities and personnel exchanges. The corporate sector is often not included in MRA-related efforts, and is a sector that is influential with Congress. Engaging the corporate sector into MRA-related efforts could help practices be picked up more widely, he suggested. Track best and unsuccessful practices on an ongoing basis, he suggested, to better determine what worked and why. He also suggested emphasizing practices that further tangible results, efficiency, and cost-effectiveness, and to clearly articulate goals and meaningful metrics to track progress of an MRA. It is invaluable to stakeholders to clearly demonstrate with understandable metrics that a decision support tool works and that science is clearly being brought into the discussion in a meaningful way, he commented. Policy makers currently view ecosystem services as a mechanism to ensure that biological and physical characteristics relate to economic valuation. There have been efforts in the past decade to educate policymakers on ecosystem services and convey the importance and value they provide; thus, he said, it is important to use consistent terminology in order for policymakers to relate what they have historically heard to what is currently being proposed.

The successful implementation of MRA will require support at all levels of an agency and within the Executive Office of the President (EOP), Dr. Schaefer said. Key agencies that he listed as needing to support MRA were the Office of Management and Budget (OMB), Council on Environmental Quality (CEQ), Office of Science and Technology Policy (OSTP), and Council of Economic Advisors (CEA). It is also important to ensure that policy and budget guidance statements from the EOP and agency leadership include clear references to MRA. Briefing key leadership charged with developing statements on policy and budget guidance helps to ensure that MRA-related concepts become part of guidance documents. Recognition in budget guidance documents helps leadership in other agencies understand that these concepts are important and should be supported. Executive and secretarial orders also help to convey the importance of the concepts.

Dr. Schaefer added that developing decision support tools in the context of regulatory mandates is essential. Agencies and policy makers are required to follow mandates so developing a tool that is consistent with mandates will further the likelihood of implementation. He warned against pursuing new regulatory mandates unless they are essential for implementation or there is strong political support for a new mandate. Additionally, he suggested the expansion of capacity budget, personnel, and programs to make use of MRA.

Multi-resource analysis is multidisciplinary; collaboration across disciplines is essential, Dr. Schaefer stated. Advancing science in support of public policy requires collaboration across the natural and social sciences, and in particular, the integration of economics with the biological and physical sciences. Dr. Schaefer said it is important to take advantage of the disciplinary strengths of agencies by collaborating across agencies. Such collaboration is invaluable given the diverse missions and supporting science and technology programs: diverse expertise, laboratories, degrees of financial support, computer and other support systems, and flexibility in making use of public-private partnerships. One avenue for collaboration, he said, is to pursue interagency and intergovernmental personnel exchanges (e.g., Intergovernmental Personnel Act Mobility Program (IPA) positions, temporary assignments, fellowships, etc.). There can also be challenges to collaborations, however, including varying degrees of policy-level support, or recognition of the value of collaboration, varying degrees of financial support, pockets of "turf protectors" at multiple levels in agencies, and limitations in sharing financial resources across agencies.

Landscape-level projects involve diverse federal, state, and local lands, private landowners, corporations, and transmission and pipeline networks; consequently, stakeholder engagement is critical, said Dr. Schaefer. There is value in multi-sector engagement early in project design and throughout the effort to pursue continuous engagement; a high level of engagement of stakeholders with diverse perspectives early helps with avoiding obstacles later. Nongovernmental organizations (NGOs) and universities should also be engaged to open access to expertise, local and regional network and knowledge, and foundations and other private-sector support. Citizen science also should be incorporated as a powerful mechanism to leverage expertise and financial support; collaborations with NGOs help to further citizen science activities. Dr. Schaefer said that stakeholder engagement is invaluable in developing decision support tools, which can be less technical in order to be useful to a broader range of users. Decision makers, he said, are often under pressure to make a decision, and want to make a decision that will be favorable to stakeholders. Decision support tools allow the decision maker to better navigate available options. Broad thinkers from multiple sectors capable of envisioning the big picture and that come together on a regular and continuous basis are important throughout a project to assess that objectives are met, appropriate metrics are chosen to benchmark progress, and stakeholders are engaged.

Gail Bingham, president emeritus of RESOLVE, presented on collaborations that bridge disciplines and perspectives. Decision making that utilizes multiple-resource analysis must be connected to the human dimension, she said. In a multi-agency collaboration, there are many different interests, expertise, and perspectives about what is in the public interest. The issues are generally complicated, and the answer is not always known. It is necessary, she said, to use everyone's ideas to help find better solutions. Collaborations are mutual efforts intended to achieve solutions that meet diverse interests by using a variety of tools and approaches (e.g., recommendations, shared decision making, and joint action). It is not a box to check or a one-size-fits-all solution, she commented.

There is a rich history of collaborations for the passage of environmental statutes in the 1960s and 1970s, Ms. Bingham said. The new environmental regulations that emerged from this period provided forums for differences to emerge as disputes among stakeholders. In response, multiple federal and state centers emerged since the 1970s. Ms. Bingham commented that she was involved in the development of the U.S. Army Corps of Engineers (USACE) Conflict Resolution & Public Participation Center of Expertise (CPCX).[1] The CPCX was created to help USACE staff anticipate, prevent, and manage water conflicts by ensuring the public interest was incorporated into USACE decision-making processes. The CPCX accomplished this by expanding the application of collaborative tools to improve water resources decision making, providing training, conducting research, and applying the application of collaborative process techniques and modeling tools to prevent and minimize water conflicts. Similarly, the Depart-

[1] Available at: www.iwr.usace.army.mil/About/TechnicalCenters/CPCXConflictResolutionPublicParticipation.aspx.

ment of the Interior's Office of Collaborative Action and Dispute Resolution (CADR) aims to improve the efficiency and effectiveness of the Department's operations, enhance communication, and strengthen relationships among all its stakeholders.[2] CADR aims to build and model conflict management competencies and integrate collaborative problem-solving and alternative dispute resolution processes in all of the Department's focus areas.

Ms. Bingham provided some examples of best practices basics. When diagnosing any particular situation or thinking about what tools to employ to respond to a particular challenge, it is important to ask about substance, process, and relationships each individually (Figure 6-1). An effective process, she said, requires attention to all three of these dimensions. It is important for the members of a stakeholder collaboration to focus on common interests and not individual positions. Other important principles include developing multiple options, using objective criteria, and developing an alternative to the collaboration. A collaboration is a shared learning process, she said. Ms. Bingham presented 10 principles from the 2008 National Research Council (NRC) report *Public Participation in Environmental Assessment and Decision Making*.[3] The principles include such basics as being clear on the purpose and being transparent and inclusive about the process (Box 6-1). The NRC committee that produced the report explicitly affirmed the 10 principles with substantial amount of evidence from the published literature.

Members of a stakeholder collaboration are seeking to be heard, and to not just have the opportunity to speak but to have their interests and ideas be valued, she said. They are also seeking to have meaningful communication and relationships, improved understanding of the issues, and to find solutions that meet their interest. Members of a collaboration want to come to an agreement with implementable results, and to do so with less stress, time, and cost. Stakeholders need to be consulted early in the process; however, Ms. Bingham cautioned that stakeholder fatigue can be a challenge to long-term collaborations. There are other challenges common to stakeholder collaborations, including adequacy of information on the issue, clarity of the decision-making process with respect to science, managing data, communication, and trust. She emphasized iteration between analysis and deliberation with a focus on decision-relevant information, explicit attention to both facts and values, explicitness about analytical assumptions and uncertainties, independent review, and reconsideration of past conclusions. Stakeholders and scientists, she said, each play an important role in these tasks; defer to stakeholders on the questions that are decision-relevant and to scientists on the information and analyses to answer those questions, she suggested. In conclusion, Ms. Bingham said to diagnose challenges early and col-

---

[2] Available at: www.doi.gov//pmb/cadr/index.cfm.

[3] National Research Council (NRC). 2008. Public Participation in Environmental Assessment and Decision Making. Washington, D.C.: The National Academies Press. [Available: http://www.nap.edu/catalog/12434/public-participation-in-environmental-assessment-and-decision-making].

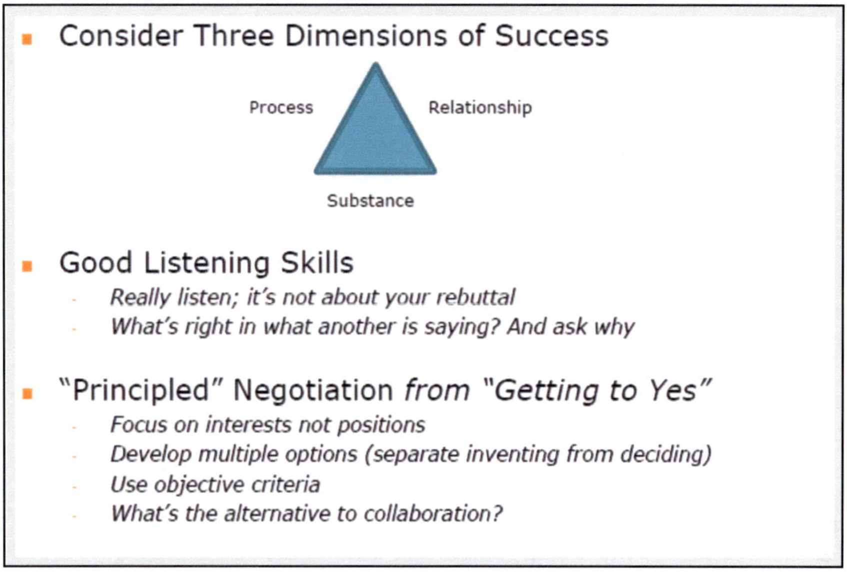

FIGURE 6-1 Best practices basics for collaboration.
SOURCE: Gail Bingham, Presentation, National Academies of Sciences, Engineering, and Medicine Workshop, June 2, 2015, Washington, D.C.

BOX 6-1
Ten Basic Principles of an Effective Stakeholder Collaboration

- Clarity of purpose (informed commitment and commitment to use the process to inform decisions)
- Timeliness in relation to decisions
- Inclusiveness (balanced, voluntary representation)
- Collaborative problem formulation and process design (group autonomy; process impartiality)
- Focus on implementation
- Accountability (good faith communication)
- Openness (transparency)
- Adequate capacity and resources
- Commitment to shared learning
- Iteration between analysis and broadly based deliberation

SOURCE: NRC, 2008.

laboratively, be inclusive, plan the process jointly, learn together, base decisions on interests (as criteria), and plan for implementation by asking if key questions were answered and if the solution is technically sound, balanced, and fair for all interests.

Paul Sandifer, professor at the College of Charleston and former chief science advisor at the National Oceanic and Atmospheric Administration (NOAA), discussed lessons learned and best practices from nearly 40 years of using science to reach and implement conservation management decision. Dr. Sandifer spent 31 years with the South Carolina Department of Natural Resources as a scientist and state fisheries administrator. As the agency administrator, Dr. Sandifer said there were a host of issues involving aquaculture; marine, coastal, and inland fisheries; conservation of special lands and biodiversity; and emergency response. One example was the ACE Basin Project, which w`as a 200,000-acre coastal conservation effort. It is considered one of the largest and most successful land conservation efforts in South Carolina.[4] The ACE Basin Project protects a vital part of the South Carolina coast between Charleston to the north and Beaufort/Hilton Head to the south, both of which are rapidly developing urban areas. The Jocassee Gorges was another example, he said, which involved the purchase of 45,000 acres of the most biodiverse area of the southern Blue Ridge Mountains.[5]

At NOAA, Dr. Sandifer was involved with the U.S. Commission on Ocean Policy, U.S. National Ocean Policy, numerous interagency working groups, and leadership roles including being co-chair of the Subcommittee on Ocean Science and Technology, the Deep Water Horizon oil spill response, and the NOAA Scientific Integrity Policy. Dr. Sandifer commented that when starting out, he felt as though fishery stakeholders viewed him as trying to take the life out of commercial fisheries. He developed four realities of management and regulation:

1. Everyone wants government to regulate somebody else but not them.
2. Perception is reality. If they view the regulator as an enforcer, then that perception needs to be altered by building trust.
3. Everyone claims they want decisions based on "best available science" but only if that science supports their point of view.
4. Facts can be interpreted and are open to interpretation by different professionals (i.e., scientists, lawyers, the public).

Dr. Sandifer also offered 10 rules of science and data used in environmental decisions:

[4] Available at: www.acebasin.net.

[5] Available at: www.dnr.sc.gov/managed/wild/jocassee/indexfull.htm.

Rule 1: Data and scientific analyses—no matter how extensive or rigorous—cannot make management and conservation decisions. They can only inform decision making.
Rule 2: There is never enough data, and information from natural and social sciences is needed along with traditional ecological knowledge.
Rule 3: Available data and analyses often do not tell you what you really need to know.
Rule 4: Data and analyses are open to different interpretations.
Rule 5: Everyone thinks they know more about the data than you, and they often have their own data. Some of that data can be from questionable sources—rigorous peer review is essential.
Rule 6: If you give scientists another week, month, year, or decade—and some funding—they will promise to answer the question you originally posed. But if you give scientists another week, month, year, or decade, they will come up with another question.
Rule 7: You will rarely have enough scientific data to unambiguously resolve a management question on the basis of science alone.
Rule 8: Nearly all of the time, you have to make decisions without defining data,
Rule 9: but that is far preferable to making a decision with no data or by ignoring data.
Rule 10: If scientists want their data to count in policy decisions, they must engage in the policy-making process. To do so, they have to learn to speak in plain language.

Dr. Sandifer commented that the stakeholder process needs to be about partnering and engaging and that one strategy is to use ad hoc committees. Putting together a committee of stakeholders, often with the most vocal opponent as the chair of the committee, is a strategy that can further issues with communication and collaboration. Only after consensus has been defined can the committee move towards finding solutions, he said. It is important to hear what stakeholders have to say and have it reflected in the output of the committee. It is also important, he said, to be prepared for vitriolic attacks, which should not be taken personally. Engaging in stakeholders and communication processes will lead to the successful implementation of multi-resource management projects.

Joseph Kiesecker from The Nature Conservancy commented that the need for a shared vision for a landscape was a concept that emerged from the panelists' presentations. He asked the panelists what capacity or expertise exists within federal agencies to shift from having a vision to engaging with the public to create a landscape-scale vision. In response, Mr. Schaefer said that federal agencies have made great strides engaging with the public on creating visions, and noted the National Environmental Protection Act, passed in 1969, contained language pertaining to participation and decision making with the public. Early on when most of these agencies were first formed they were overwhelmed and had limited resources, and were not able to effectively engage with the public. Many agencies now have in-house entities to advance collaboration and conflict resolution. Federal agencies still need to find ways to more effectively engage stakeholders. Dr. Sandifer commented that a key shortfall is the lack of training capacity for scientists within federal agencies for public engagement. Most expertise in this field lies in the private sector and NGOs.

A participant provided an anecdote about a U.S. Army Corps of Engineer colleague who mediated parties as a central component to his career. The colleague would request information from whomever they were supporting on a project prior to meeting with them, and was able to use that information during the mediation process. The colleague found that overall the success of a decision support tool or model was largely dependent on the audience's willingness to accept and work with the tool. Ms. Bingham replied that different terms like conflict or situation assessment have been used to describe the human dynamic of a mediation situation. Sometimes, she said, when an agreement cannot be reached, it is not due to the decision process. An example Ms. Bingham provided from Maryland related to a disagreement over Clean Water Act funding allocation. Scientists upstream related nutrient issues to phosphorus loading, but those downstream related the issues to nitrogen loading. The disagreement really was over the allocation of funding to point versus non-point source contributors of nutrients to the Chesapeake Bay. A mediator was able to work with stakeholders to find which disagreements were decision-relevant to implement land-use plans and ultimately to negotiate funding allocations. Stakeholders need to be reminded about what the decision is to make, how it is going to be implemented, and who will be affected by the implementation, she pointed out.

Dr. Bartuska commented that despite all the discussion about analysis and modeling, the decision-making process is determined by the human factors of communication, social engagement, and participation. An ever-persistent challenge to governmental collaborations is that as budgets are cut, such efforts are often the first programs to be cut, she said. Because agencies are always funding-limited, that training gap is intractable; however, some agency employees have such skills already which could be cultivated and provided opportunity to function in a more collaborative role.

Dr. Sandifer replied that one way to work around the funding challenge is to work with university and NGO partners. This would allow for distributed activities that may relate back to congressional districts and gain more political traction for support. It would also provide an extra benefit of collaboration building among people from different backgrounds and agencies who would end up having similar training. Pushing the envelope on science application, he said, would also find support among philanthropic groups who can bridge funding gaps for start-up training-related activities.

CONCLUDING REMARKS

Ingrid Burke facilitated a recap of the workshop by describing themes that she identified from speaker presentations and panel discussions. The social context, for example, is critical for multiple resource assessments. It is must be clear who the decision maker is in a multi-stakeholder situation in order to appropriately match the scale of the science to the scale of the decisions being made. Decision makers, however, often have variable jurisdictions and the scale of the problems can change. Dr. Burke said that the optimization of multiple resources must occur where the entire landscape is optimized and not by focusing on individual locations or "pixels"—a term several speakers used.

Dr. Burke emphasized that a key lesson learned from several of the speakers' presentations was that multiple, large-scale projects can lead to stakeholder fatigue. These projects often require the most analysis and also have major conflicts. Another challenge to multi-resource assessments is a sudden change in priorities for a given area with lots of investments. These challenges to the stakeholder process require that uncertainty be communicated well and at the right level of detail. At times, she said, details are very important, but there are also times that communicating at a conceptual level is important—there is a balance between providing simplicity and accuracy. Communication training and development for scientific professionals is important. There are several university programs that provide this training to emerging scientists, but providing it to mid- and senior-level career scientists should also be encouraged.

Structured decision analysis, she said, is important not only for outcomes, but also as a way of communicating with stakeholders. Landscape planning and ecosystem services analysis often do not fully take into consideration the step of addressing who benefits from the analysis—demographic results should be included into the analyses of ecosystem services. A recurring theme from speakers was that conducting stakeholder engagement at the beginning of the analysis is critical. Stakeholder engagement early in the process allows modelers and scientists to learn more about each other priorities.

Dr. Burke said that multi-resource assessments are conceptually easy to conduct, but empirically difficult to execute. The answer is not always clear with the available data and analyses, and decisions often need to be made in the absence of exhaustive databases. How to manage data is also a recurring challenge. There are nonlinear variables that can be aggregated in an analysis, but other data, such as value data, that are difficult to aggregate. Managing and aggregating data appropriately will be critical to successful analyses. Decision frameworks also change because of ongoing processes of developing science and decision tools. It is important, she said, to understand that decision frameworks continually change as more information is developed. Interagency collaborations also have many challenges that limit their efficacy, including sharing budget and personnel. Dr. Burke concluded that it may be useful to develop best practices for decision analysis tools and multi-criteria or multi-resource assessments to ensure consistency in results.

Appendix A

Workshop Agenda

The National Academies of Sciences, Engineering, and Medicine
Room 120
2101 Constitution Ave., NW
Washington, D.C.

June 2, 2015

8:30 AM **Welcome**
Jerry Miller, Director, Science and Technology for Sustainability Program, The National Academies of Science, Engineering, and Medicine

8:35 AM **Introduction**
Ingrid Burke, Director, Haub School of Environment and Natural Resources, University of Wyoming

8:40 AM **Keynote Panel: Decision-Oriented Approaches to Natural Resource Management**
Suzette Kimball, Acting Director, U.S. Geological Survey
Steven Ellis, Deputy Director, Bureau of Land Management
Ann Bartuska, Deputy Under Secretary for Research, Education, and Economics, U.S. Department of Agriculture

9:40 AM **Identifying Needs and Challenges for Landscape and Multi-Resource Analyses**
Moderator: Kit Muller, Management and Program Analyst, Bureau of Land Management
Elsa Haubold, National Landscape Conservation Cooperatives Coordinator, U.S. Fish and Wildlife Service
Ione Taylor, Executive Director of Earth and Energy Resources Leadership, Queens University

10:50 AM BREAK

11:05 AM **Methods for Spatial Analysis: Identifying Scenarios**
Moderator: Steve Bergman, Principal Regional Geologist, Shell International Exploration and Production Co.
Lisa Wainger, Research Professor, University of Maryland
Murray Hitzman, Charles F. Fogarty Professor Economic Geology, Colorado School of Mines
Dean Urban, Professor of Landscape Ecology, Duke University

12:15 PM LUNCH

1:00 PM **Methods for Evaluating Scenarios: Reconciling Quantities and Values**
Moderator: Patrick Huber, Project Scientist, Information Center for the Environment, University of California, Davis
Richard Bernknopf, Research Professor, University of New Mexico and USGS Net Resource Assessment
Karen Jenni, President, Insight Decisions LLC
Robert Johnston, Director and Research Professor, George Perkins Marsh Institute, Clark University

2:10 PM **Discussion**

3:00 PM BREAK

3:15 PM **Multi-Disciplinary and Cross-Agency Synthesis**
Moderator: Joseph Kiesecker, Lead Scientist, The Nature Conservancy's Conservation Lands Team
Mark Schaefer, Global Fellow, Science and Technology Innovation Program, Wilson Center
Gail Bingham, President Emeritus, RESOLVE
Paul Sandifer, College of Charleston and Former Chief Science Advisor, National Oceanic and Atmospheric Administration

4:25 PM **Discussion**

4:55 PM **Wrap-up**

5:00 PM ADJOURN

Appendix B

Workshop Participants

National Academy of Sciences
Washington, DC
June 2, 2015

Ingrid Burke (Chair)
University of Wyoming

Greg Arthaud
U.S. Forest Service

Ann Baker
U.S. Army Corps of Engineers

Ann Bartuska
U.S. Department of Agriculture

Steve Bergman
Shell International Exploration and Production Company

Richard Bernknopf
University of New Mexico

Gail Bingham
RESOLVE

James Boyd
Resources for the Future & National Socio-Environmental Synthesis Center

Patricia Bright
U.S. Geological Survey

David Brookshire
University of New Mexico, Albuquerque

Frank Casey
U.S. Geological Survey

Janet Cushing
U.S. Geological Survey

Jay Diffendorfer
U.S. Geological Survey

Shasta Ferranto
Bureau of Land Management

Sarah Gerould
U.S. Geological Survey

Debra Gilliam
Caset Associates, Ltd.

Steven Ellis
Bureau of Land Management

Seth Haines
U.S. Geological Survey

Jaelith Hall-Rivera
Office of Management and Budget

Tomer Hasson
U.S. Department of the Interior

Elsa Haubold
U.S. Fish and Wildlife Service

Murray Hitzman
Colorado School of Mines

Patrick Huber
University of California, Davis

Karen Jenni
Insight Decisions LLC

Robert Johnston
Clark University

Emi Kameyama
The National Academies of Science, Engineering, and Medicine

Joseph Kiesecker
The Nature Conservancy

Suzette Kimball
U.S. Geological Survey

Jon Kolak
U.S. Geological Survey

Jeffrey Krause
U.S. Army Corps of Engineers

Sophia B. Liu
U.S. Geological Survey

Edward Maillett
U.S. Fish and Wildlife Service

Joe Manous
U.S. Army Corps of Engineers

Larry Meinert
U.S. Geological Survey

Jerry Miller
The National Academies of Science, Engineering, and Medicine

Kit Muller
Bureau of Land Management

Anne Neale
U.S. Environmental Protection Agency

Vito Nuccio
U.S. Geological Survey

David Olson
U.S. Army Corps of Engineers

John Organ
U.S. Geological Survey

Emily Pindilli
U.S. Geological Survey

Yasmin Romitti
The National Academies of Science, Engineering, and Medicine

Paul Sandifer
College of Charleston

Jennifer Saunders
The National Academies of Science, Engineering, and Medicine

Mark Schaefer
Wilson Center

Rudy Schuster
U.S. Geological Survey

Carl Shapiro
U.S. Geological Survey

David Simpson
U.S. Environmental Protection Agency

Ione Taylor
Queens University

Dean Urban
Duke University

Lisa Wainger
University of Maryland

Rob Winthrop
Bureau of Land Management

Paul Young
U.S. Geological Survey

Appendix C

Biographies of Planning Committee, Speakers, and Staff

INGRID C. BURKE (Planning Committee Chair) is director of the Haub School of Environment and Natural Resources of the University of Wyoming and of its Ruckelshaus Institute. She also is a professor and holds a Wyoming Excellence Chair in the Department of Botany and the Department of Ecosystem Science and Management. She is a former professor and University Distinguished Teaching Scholar in the Warner College of Natural Resources of Colorado State University. Dr. Burke is an ecosystem scientist and has particular expertise in carbon and nitrogen cycling of semiarid ecosystems. She directed the Shortgrass Steppe Long Term Ecological Research team for six years and other large interdisciplinary research teams funded by the National Science Foundation, the Environmental Protection Agency, the National Aeronautics and Space Administration, and the National Institutes of Health. She was designated a US Presidential Faculty Fellow, has served on the National Academies of Sciences, Engineering, and Medicine Board on Environmental Studies and Toxicology, and was a member of the National Academies of Sciences, Engineering, and Medicine's Committee on Scientific Tools and Approaches for Sustainability. She also served as co-chair of the Committee on Economic and Environmental Impacts of Increasing Biofuels Production. She was recently elected a Fellow of the American Association for the Advancement of Science. Dr. Burke received a Ph.D. in botany from the University of Wyoming.

ANN BARTUSKA is deputy undersecretary for research, education, and economics (REE) in the U.S. Department of Agriculture (USDA). Previously, she was deputy chief for research & development in the USDA Forest Service, a position she held since January 2004. She recently served as acting USDA deputy undersecretary for natural resources and environment from January to October of 2009, and was the executive director of the Invasive Species Initiative in the Nature Conservancy. Prior to this, she was the director of the Forest and Rangelands staff in the Forest Service in Washington, DC. She co-chaired the Academies' Roundtable on Science and Technology for Sustainability from 2010-2013. She currently co-chairs the Ecological Systems subcommittee of the Committee on Environment and Natural Resources of the White House National Science and Technology Council. She is active in the Ecological Society of America, serving as vice-president for public affairs from 1996-1999 and as president from 2002-2003. She has served on the board of the Council of Science Society Presidents and is a member of AAAS and of the Society of American Foresters. She is an ecosystem ecologist with degrees from Wilkes College (B.S.), Ohio University (M.S.), and West Virginia University (Ph.D.).

STEVE BERGMAN (Planning Committee Member) is a principal regional geologist in the Global Geology Upstream Exploration Research team at Shell International Exploration & Production Co. (Houston), and adjunct professor with Southern Methodist University (SMU, Dallas). Prior to Shell, he was principal research geologist with ARCO R&D for 20 years in Dallas and a visiting scholar at Bullard Laboratories, Cambridge University in 1996-1997. Dr. Bergman is exploration research geologist and geoscience educator with over 30 years of industry experience applying unconventional integrated field and laboratory approaches (completing over one hundred worldwide minerals and petroleum exploration and production projects) and five years of university teaching at UT Dallas and SMU. Dr. Bergman is a world-class expert in tectonics, regional structure, field geology, basin analysis, hard rock petrology, volcanology, and geochronology with 30 months of field expeditions in 17 U.S. states and 22 countries. He is co-author of a textbook (*Petrology of Lamproites*, Plenum Press, New York), over 40 journal papers, 120 internal company reports, and 150 conference and seminar presentations. He has served on the National Academies of Sciences, Engineering, and Medicine Roundtable on Science and Technology for Sustainability (STS) since 2012, the NRC STS Landscape Analysis Committee in 2015, and as an advisor to the U.S. Department of State on Arctic Geology Matters. He is a Fellow of the Geological Society of London. Dr. Bergman earned his M.A. and Ph.D. in geology from Princeton University (1979 and 1982) and his B.S. in geology from University of Dayton (1977).

RICHARD L. BERNKNOPF has been a research professor in the Department of Economics at the University of New Mexico (UNM) since 2011. Before joining the faculty at UNM, Dr. Bernknopf was an economist with the U.S. Geological Survey (USGS) for more than 38 years. His research focuses on the demonstration of the relevance (value to society) of scientific data including earth observation and the translation of that information into a form compatible with decision-making processes. During his tenure at USGS, he was a consulting professor and co-director of the Center for Earth Science Information Research at Stanford University and co-director of the Spatial Integration Laboratory for Urban Systems at the University of Pennsylvania. Currently he is an associate with the Science Impact Laboratory for Policy and Economics at the University of New Mexico and the Wharton Geospatial Initiative at the University of Pennsylvania.

GAIL BINGHAM is president emeritus of RESOLVE. She has mediated environmental, natural resources, community planning, and public health issues on a full-time basis since the late 1970s. She is a nationally recognized pioneer in promoting consensus-building tools in public decision making and is the 2006 winner of the Mary Parker Follett Award of the Association for Conflict Resolution. Ms. Bingham works with leaders at the highest levels of government and the private sector, and has served as a mediator for a wide variety of federal, state, and local agencies and private parties on such diverse subjects as sustainable water management, endangered species, wetlands policy, allocation of water rights, drinking water regulations, funding infrastructure costs for water and wastewater utilities, groundwater protection, ocean and coastal management issues, hydro-electric relicensing, chemicals policy, solid waste source reduction, hazardous waste management, oil spill contingency plans, pesticides policy, children's environmental health, and local community land use and infrastructure issues. She currently serves on advisory committees for the Haub School of Environment and Natural Resources at the University of Wyoming and for the Center for Environmental Policy at American University. She also served on the Panel on Public Participation in Environmental Assessment and Decision Making for the National Academy of Sciences. She served two terms on the Board of Directors of the Society of Professionals in Dispute Resolution (now ACR); was the founding chair of its environment and public policy sector and a president of its Washington DC Chapter; and served on numerous committees, including the first and third Commissions on Mediator Qualifications. She also has testified before Congress on several occasions, on topics such as the Administrative Dispute Resolution Act and the Negotiated Rulemaking Act.

DOMINIC A. BROSE (NRC Staff) is a program officer for the Science and Technology for Sustainability Program (STS) at the National Academies of Science, Engineering, and Medicine. Prior to STS, Dr. Brose was a research assistant with the National Academy of Medicine where he collaborated on science policy reports sponsored by the Department of Veteran Affairs (VA) addressing the potential for adverse health effects from exposure of select military personnel to environmental contaminants. Previously, he was an environmental scientist at ToxServices LLC, where he reviewed product formulations for EPA's Design for the Environment (DfE) program, a third-party service that evaluated product formulations against human health and environmental screening criteria. Dr. Brose received his M.S. and Ph.D. in environmental soil chemistry from the University of Maryland, and his B.S. in natural resources and environmental science from Purdue University.

STEVE ELLIS is the deputy director for operations at the Bureau of Land Management (BLM). As a veteran land manager, Mr. Ellis has spent 21 years as a line officer in the BLM and the U.S. Forest Service, and as BLM-Idaho state director. Among other Forest Service positions, he was the forest supervisor for the Wallowa-Whitman National Forest in Oregon. He has also held numerous jobs throughout the BLM, including acting associate district manager in Las Vegas, district manager of Oregon's Lakeview District, and acting deputy state director in Alaska. He has extensive wildfire experience in the West, including serving as an incident commander, and has worked as a budget officer in the agency's headquarters office in Washington, D.C. During his time in Washington, Mr. Ellis was selected and served as a congressional fellow in the U.S. Senate. He holds a B.S. in forestry from Southern Illinois University at Carbondale and an M.S. in geography from Northern Illinois University at DeKalb.

ELSA HAUBOLD has served as the national Landscape Conservation Cooperative coordinator since September 2013. Previously, she worked for 12 years on wildlife diversity and endangered species issues at the state, regional, and national level with the Florida Fish and Wildlife Conservation Commission. Dr. Haubold also has a wealth of nongovernmental organizational experience, having coordinated the Texas Marine Mammal Stranding Network. Dr. Haubold is passionate about working with partners and stakeholders to find common ground and solutions to seemingly insurmountable conservation challenges. She has a B.S. in wildlife and fisheries science, M.S. in veterinary anatomy from Texas A&M University, Ph.D. in pathology from the University of Texas Medical Branch, and MBA from the University of Houston Clear Lake.

MURRAY W. HITZMAN worked in the petroleum and minerals industries from 1976-1993, primarily doing mineral exploration worldwide and was largely responsible for Chevron Corporation's Lisheen Zn-Pb-Ag deposit discovery in Ireland (1990). Dr. Hitzman served in Washington, D.C., as a policy analyst in the U.S. Senate for Senator Joseph Lieberman (1993-1994) and in the White House Office of Science and Technology Policy (1994-1996). In 1996 he was named the Fogarty Professor in Economic Geology at the Colorado School of Mines and served as head of the Department of Geology and Geological Engineering from 2002-2007. While his research in economic geology with graduate students has been conducted around the world, for the past 17 years he has focused his attention on the Central African Copperbelt. Dr. Hitzman has published extensively on the geology and geochemistry of mineral deposits and on natural resource policy issues. He served as the president of the Society of Economic Geologists in 2006. He is a member of the National Academies of Sciences, Engineering, and Medicine Committee on Geological and Geotechnical Engineering. He has also served on the boards of a number of junior mineral exploration and mining companies. He holds B.A. degrees in geology and anthropology from Dartmouth, an M.S. in geology from the University of Washington, and a Ph.D. in geology from Stanford.

PATRICK HUBER has held the positions of postdoctoral scholar and project scientist at the Information Center for the Environment (ICE) at the University of California, Davis. As a conservation scientist, he has performed spatial and other analyses for a variety of projects both in California and at the global scale. He has worked with a wide range of collaborators in public agencies, private organizations, universities, and corporations. Some areas of emphasis in his work include landscape connectivity; reserve design; conservation prioritization; mitigation for infrastructure impacts; and geospatial analysis (primarily using GIS). Currently much of his work focuses on integrating stakeholders and conservation priorities at the regional scale to help produce systematic and integrated conservation plans. Some of his primary tools include Marxan reserve selection software and GIS connectivity modeling tools. He received his Ph.D. in geography from the University of California, Davis

KAREN JENNI is the founder and president of Insight Decisions, LLC. She is a professional decision analyst with over 25 years of consulting experience leading, managing, and conducting projects involving the application of decision analysis and risk management techniques to large-scale energy and environmental policy problems. Her areas of expertise include decision analysis, multi-attribute analysis, behavioral decision theory, and integrated risk modeling. She has been involved in several previous Academies' studies as a consultant, committee member, and as a reviewer. She received a Ph.D. in engineering and public policy from Carnegie Mellon University.

ROBERT J. JOHNSTON (Planning Committee Member) is director of the George Perkins Marsh Institute and professor of economics at Clark University. Dr. Johnston's research addresses benefit cost analysis, economic valuation, benefit transfer, and ecosystem services, with an emphasis on aquatic, riparian, and coastal systems. His recent work has focused on the economics of coastal vulnerability and adaptation. He is a current member of the U.S. EPA Science Advisory Board, the Ecosystem Science and Management Working Group of the NOAA Scientific Advisory Board, the Management Committee and Science Advisory Board of the Narragansett Bay Estuary Program, the Senior Advisory Board of the Connecticut Sea Grant Program, the Program Advisory Council of the New York Sea Grant Program, and the Program Committee for the Charles Darwin

Foundation in Galapagos, Ecuador. He also serves on the Council on Food, Agricultural and Resource Economics (C-FARE) Blue Ribbon Panel on Natural Resources and Environmental Issues. He has published over 100 peer-reviewed articles and chapters on the economics of the environment and natural resource management. He has a Ph.D. in resource economics from the University of Rhode Island and a B.A. in economics from Williams College.

JOSEPH KIESECKER (Planning Committee Member) is a lead scientist for The Nature Conservancy's Conservation Lands Team. In this capacity, his main responsibilities include developing new tools, methods, and techniques that improve conservation. He pioneered the Conservancy's Development by Design strategy to improve impact mitigation through the incorporation of predictive modeling to provide solutions that benefits conservation goals and development. He also conducts his own research in areas ranging from disease ecology to the effectiveness of new conservation tools such as conservation easements. He has held faculty appointments at Yale University, Penn State University, and currently holds a faculty appointment at the University of Wyoming. He has been a Donnelly Fellow and has received funding for his research from National Institutes of Health, the National Science Foundation, the International Union for Conservation of Nature (IUCN), and numerous private foundations. Dr. Kiesecker has published over 100 articles, on topics ranging from climate change to the effectiveness of conservation strategies; examples of his work have been published in Nature, Proceedings of the National Academy of Sciences, Conservation Biology, Ecology and American Scientist. His training was in ecology, conservation biology, and animal behavior, with a Ph.D. from Oregon State University in 1997.

SUZETTE KIMBALL is acting director of the U.S. Geological Survey (USGS), and is responsible for leading the nation's largest water, earth, and biological science, and civilian mapping agency. Prior to becoming the acting director, Dr. Kimball was the USGS deputy director. In 2008, she became the acting associate director for geology, and prior to that was the director of the USGS eastern region, starting in 2004. She joined the USGS as eastern regional executive for biology. In that position, she built many partnerships, helped shape programs, and led the establishment of the USGS Florida Integrated Science Center. She came to the USGS from the National Park Service in Atlanta, where she was associate regional director. She entered the National Park Service as a research coordinator in the Global Climate Change Program, became southeast regional chief scientist, and then associate regional director. She was assistant professor of environmental sciences at the University of Virginia, co-director of the Center for Coastal Management and Policy and marine scientist at the Virginia Institute of Marine Science, and she managed coastal morphology and barrier island studies in the U.S. Army Corps of Engineers. Dr. Kimball has a doctorate in environmental sciences with a specialty in coastal processes from the University of Virginia, a master's in geology and geophysics from Ball State University, and a bachelor's in English and geology from the College of William & Mary.

JERRY L. MILLER (NRC Staff) has been director of the Science and Technology for Sustainability (STS) Program since February 2, 2015. A senior executive with expertise in science and resource management policy with more than 20 years of experience, Dr. Miller is the NRC's senior scientist driving policy and program direction on sustainability-related issues. Previously, Dr. Miller served as president of Science for Decisions, a consulting practice that he founded to ensure that solid science is available to inform policy and management decisions that impact natural resources and the livelihoods that depend upon them. From 2009 until 2013, Dr. Miller served as assistant director for ocean sciences at the White House Office of Science and Technology Policy (OSTP). During his time at OSTP, Dr. Miller was instrumental in the creation of the nation's first National Ocean Policy and the development of its foundational science priorities. He was founding co-director of the National Ocean Council Office and later served as its deputy director for Science and Technology. Before taking on his role at OSTP, Dr. Miller was technical director and director of research at the Consortium for Oceanographic Research and Education (now the Consortium for Ocean Leadership), where he had management and oversight responsibilities for the program offices of the U.S. National Oceanographic Partnership Program, the national and international Census of Marine Life programs, and other community-wide activities. As associate director for ocean, atmosphere, and space sciences at the Office of Naval Research's global office in London, he built international programs in ocean and atmosphere modeling as well as remote sensing. Dr. Miller has published widely in the peer-reviewed literature and has made significant contributions to several major federal policy documents. His work has been recognized with awards both in the United States and abroad, including a Distinguished Career Achievement Award from the University of Rhode Island. Dr. Miller received his B.S. in marine science from the University of South Carolina, M.S. in oceanography from University of Rhode Island, and Ph.D. in meteorology and physical oceanography from University of Miami.

KIT MULLER serves as management and program analyst at the U.S. Department of the Interior (DOI) Bureau of Land Management (BLM). As a strategic planner, Mr. Muller has guided the BLM's efforts to systematically understand and address the effects of climate change, wildfire, invasive species, industrial development, and urban growth on landscape management. He received his B.S. in social anthropology from Harvard University and M.S. in public policy from the University of California, Berkeley.

ELIZABETH MURRAY (Planning Committee Member) has worked as a wetland scientist for 20 years, specializing in wetland assessment, ecological restoration, and resource management. Ms. Murray currently works as a research biologist in the Wetlands and Coastal Ecology branch of the Environmental Laboratory (EL) of the U.S. Army Engineer Research and Development Center (ERDC). She has co-authored eight Hydrogeomorphic (HGM) functional assessment regional guidebooks for wetlands in various parts of the country, covering over 40 regional subclasses. She has also developed spreadsheet Functional Capacity Index calculators and interactive data forms, as well as scientific illustrations, for many others. She has performed wetland functional assessments on large civil works projects. Although she is most involved in HGM assessment approaches, she has also researched, helped develop, or reviewed several other wetland assessment methods, including California Rapid Assessment Method (CRAM), Landscape Development Intensity (LDI) index approaches, and remote sensing techniques. She is co-author on one of the resulting atlases, as well as a Wetlands article on the techniques used in the mapping. She is co-PI on the Corps' national effort to investigate the potential for incorporating ecosystem goods and services in civil works and restoration planning. Ms. Murray has a B.S. in biology from the University of California, San Diego, and M.S. in environmental science from the School of Public and Environmental Affairs, Indiana University.

PAUL SANDIFER, recently retired from NOAA, is currently a part-time research associate (professor) in the School of Sciences and Mathematics at the College (University) of Charleston, South Carolina where he conducts research and advises graduate students. Dr. Sandifer is a coastal ecologist with a broad background in research, natural resource management, science policy, and the intersection of marine ecosystem health and human health. His prior career includes nearly 12 years in NOAA, where he served as a senior scientist, science advisor to the NOAA administrator, and chief science advisor for NOAA's National Ocean Service, and he was responsible for development of NOAA's Oceans and Human Health Program. Before NOAA, he worked 31 years with the South Carolina Department of Natural Resources where he served in a variety of science and management positions including as agency director. Recently, he led establishment of a NOAA-wide effort in ecological forecasting, initial stages of development of NOAA's RESTORE Act Science Program for the Gulf of Mexico, creation of NOAA's highly regarded scientific integrity policy, and played key roles in the development of the National Ocean Policy. He has served on numerous boards and interagency committees, including the U.S. Commission on Ocean Policy, the NAS Marine Board and several Roundtables, the CENRS Subcommittee on Ocean Science and Technology, the U.S. National Committee for the Census of Marine Life, and the Founding Board of Directors of the South Carolina Aquarium.

MARK SCHAEFER is presently a global fellow affiliated with the Woodrow Wilson International Center for Scholars. He previously served as assistant secretary of commerce for conservation and management, and deputy administrator of the National Oceanic and Atmospheric Administration, responsible for coastal and ocean programs. From 2008-2013, he was director of the U.S. Institute for Environmental Conflict Resolution at the Morris K. Udall and Stewart L. Udall Foundation, a federal agency. From 2007-2008, he served as a senior advisor and consultant to several organizations on environmental science and technology policy, including the Woodrow Wilson Center. From 2006-2007, he was CEO of the Global Environment and Technology Foundation, an organization dedicated to the advancement of sustainable development and environmental technologies worldwide. From 2000-2006, he was president and CEO of NatureServe, an international nonprofit scientific organization providing information and analytical tools to inform conservation decision making. From 1996-2000, he served as deputy assistant secretary, and later acting assistant secretary of the Interior for Water and Science. In this position he provided policy guidance to the U.S. Geological Survey and the U.S. Bureau of Reclamation. Dr. Schaefer was acting director of the U.S. Geological Survey from October 1997 to February 1998. He previously served for three years as assistant director for environment in the White House Office of Science and Technology Policy, where he was responsible for a variety of energy and environmental science, technology, and education issues, including a major initiative to advance the development and diffusion of environmental and renewable energy technologies nationally and globally. He served two terms as a member of the Board on Earth Sciences and Resources of the National Academies of Sciences, Engineering, and Medicine, and was a member of the Commission on Education and Communication of the World Conservation Union (IUCN). A biologist by training, he received a B.A. (zoology and botany) from the University of Washington and Ph.D. (neurosciences) from Stanford University.

IONE TAYLOR joined Queen's University in 2014 as executive director, Earth and Energy Resources Leadership. She is responsible for developing a graduate-level professional program that integrates geosciences, engineering, economics, legal, societal, and policy concepts into a curriculum to develop the next generation of enterprise leaders for the natural resource sector. She began her career in the petroleum industry, working as an operations geologist on drilling wells in the Gulf of Mexico. She spent the next 15 years focused on domestic and international hydrocarbon exploration, holding multiple scientific and technical positions at Amoco Production Company and British Petroleum. Dr. Taylor eventually moved into senior leadership positions including R&D director of worldwide technology applications, vice president of overseas exploration, and upstream technology group lead for Worldwide Reservoir Description. She subsequently entered U.S. federal government service with the U.S. Geological Survey (USGS) in the Department of the Interior, with positions focused on energy and mineral resource security for the United States, interdisciplinary environmental science, and application of satellite remote sensing for earth observation. Most recently, as USGS associate director for energy and minerals, and environmental health, Dr. Taylor served as the senior executive responsible for oversight of geological research and assessment programs for energy and mineral resources and economics to inform natural resource management. She holds a B.S. degree in chemistry and M.S. and Ph.D. degrees in geology, is a graduate of Thunderbird School of Global Management, and holds an executive certificate in Strategy and Innovation from Sloan School of Management at Massachusetts Institute of Technology.

DEAN URBAN is a professor of landscape ecology at Duke University's Nicholas School of the Environment. Dr. Urban's interest in landscape ecology focuses on the agents and implications of pattern in forested landscapes. Increasingly, his research is in what has been termed "theoretical applied ecology," developing new analytic approaches to applications of immediate practical concern such as conservation planning. A hallmark of his lab is the integration of field studies, spatial analysis, and simulation modeling in extrapolating fine-scale empirical understanding of environmental issues to the larger space and time scales of management and policy. Dr. Urban received his Ph.D. from University of Tennessee, Knoxville, his M.A. from Southern Illinois University, and his B.A. from Southern Illinois University.

LISA A. WAINGER is a research professor of environmental economics at the University of Maryland Center for Environmental Science. Her primary research interest is in developing integrated ecological and economic analysis tools to analyze risk and economic efficiency of policy alternatives. Her work emphasizes the spatial variability of ecosystem service benefits to support decisions for prioritizing restoration and preservation. She has over 20 years of experience working, nationally and internationally, with government agencies, nonprofit organizations, and private firms to address issues of agro-ecosystem management, invasive species, wetland mitigation, preserving habitat for rare species, and water quality. She currently serves or has served on numerous federal and other advisory boards including the Scientific and Technical Advisory Committee to the Chesapeake Bay Program, White House Council on Environmental Quality, and the National Fish and Wildlife Foundation. She received her B.S. in earth science from the University of California, Santa Cruz, and her Ph.D. in environmental and ecological economics at the University of Maryland, College Park.